COMPETITIVE & GREEN

SUSTAINABLE PERFORMANCE IN THE ENVIRONMENTAL AGE

Dennis C. Kinlaw, Ed.D.

PHILA COLLEGE PHARMACY & SCIENCE
J.W. ENGLAND LIBRARY
4200 WOODLAND AVENUE
PHILADELPHIA, PA 19104-4491

Amsterdam • Johannesburg • London
San Diego • Sydney • Toronto

Copyright © 1993 by Pfeiffer & Company

Copyright under International, Pan American, and Universal Copyright Conventions. All rights reserved. No part of this book may be reproduced or transmitted in any form or by any means, electronic or mechanical, including photocopying, recording, or by any information storage-and-retrieval system, without written permission from the publisher. Brief passages (not to exceed 1,000 words) may be quoted for reviews.

This publication is designed to provide accurate and authoritative information in regard to the subject matter covered. It is sold with the understanding that the publisher is not engaged in rendering legal, accounting, or other professional service. If legal advice or other expert assistance is required, the services of a competent professional person should be sought. *From a Declaration of Principles jointly adopted by a Committee of the American Bar Association and a Committee of Publishers.*

Editor: Arlette C. Ballew
Interior Designer: Lee Ann Hubbard
Cover: John Odam Design Associates

Pfeiffer & Company
8517 Production Avenue
San Diego, CA 92121-2280

ISBN: 0-89384-227-3

Library of Congress Catalog Card Number: 92-51018

Printed in the United States of America.

Printing 1 2 3 4 5 6 7 8 9 10

This book is printed on acid-free, recycled stock that meets or exceeds the minimum GPO and EPA specifications for recycled paper.

DEDICATION

This book is dedicated to all the children in my family and all those in yours.

TABLE OF CONTENTS

Preface	xiii
Introduction	1
Competitive & Green	3
The Need for an Integrating and Organizing Concept	4
Sustainable Performance	6
Purposes of the Book	8
General Description and Overview of Content	9
1 The Sustainable-Performance Management Model	**13**
Examples of Sustainable Performance	14
The Sustainable-Performance Management Model	19
Uses of the Model	21
Description of the Model	22
Summary	34
2 Milestones for Sustainable Performance	**37**
Derivation of Milestones	38
Qualitative Characteristics of Management	39
The Milestones for Sustainable Performance	47
Summary	65

3	**Pressures to Respond**	**67**
	Our Slow Response to the Global Threat	68
	The Problems Are Systemic and Global	71
	The Changes and Issues	73
	Growing Pressures to Change	74
	Compliance	77
	Punitive Fines and Costs	79
	Personal Culpability and Imprisonment	81
	Environmental Activist Organizations	83
	An Aroused Citizenry	85
	Societies, Coalitions, and Associations	89
	International Codes for Environmental Performance	94
	Environmentally Conscious Investors	96
	Consumer Preference	99
	Global Markets	105
	Global Politics and International Organizations	107
	Competition	108
	Other Pressures	112
	Summary	115
4	**Characteristics of and a Systems Model for Sustainable Performance**	**119**
	Sustainable Development and Sustainable Performance	120
	Characteristics of Sustainable Performance	127
	A Systems Model for Sustainable Performance	131
	Summary	151

5	**The Principles of Sustainable Performance**	**155**
	Principles of Sustainable Performance	156
	Summary	187
6	**Response Levels**	**191**
	The Response Model for Sustainable Performance	192
	Pressures	196
	Screens	196
	Motives	203
	Competencies	210
	Levels of Response	213
	Summary	220
7	**Strategies and Tools**	**223**
	Section I: The General Strategies	224
	Strategy One: Practicing Conservation and Paying Attention to Details	225
	Strategy Two: Modifying or Replacing Existing Processes, Products, and Services	227
	Strategy Three: Reclaiming Waste and Secondary Products	228
	Strategy Four: Reducing the Use of Materials	231
	Strategy Five: Finding New, Green Market Niches	233
	Summary of Section I	235
	Section II: Tools for SP	235
	The Sustainable-Performance Assessment (SPA)	236
	Audits	239

	Auditing Programs	239
	Developing an Auditing Program	240
	Environmental Audits	252
	The Auditing Process	255
	Benchmarking	261
	Life-Cycle Analysis	264
	Summary	268
8	**Putting It All Together**	**271**
	The New Rules Are Green	271
	Playing by the Green Rules	277
	What Leaders Can Do	278
	Conclusion	281
Appendix		**285**
	Section I: International Environmental Business Codes	*286*
	CERES	286
	International Chamber of Commerce	289
	Section II: Resources for Sustainable Performance	*293*
	International Resources	294
	U.S. Resources	295
	Section III: Environmental Laws	*297*
	U.S. Laws	298
	International Laws	301
	Section IV: Journals and Magazines	*302*
	Section V: Environmental-Audit Checklists	*304*

Water Pollution Checklist	305
Checklist to Reduce Energy Consumption of Company Vehicles	306
Checklist to Assess Risk in Land or Facilities Purchase	307
Section VI: The Sustainable-Performance Assessment (SPA)	*308*
Response Levels	308
Uses of the SPA	311
Directions for Using the SPA	312
Interpreting the SPA	313
Company/Organization Index	**323**
Subject Index	**329**

ACKNOWLEDGMENTS

The general topic of this book was suggested to me over two years ago by Dick Roe, vice president for acquisitions at Pfeiffer & Company. I thank Dick for the idea and for the confidence that he expressed in my ability to produce the book.

Many more people have contributed to this book than to any of the others that I have written. David Duperault served as my research assistant for a summer and discovered leads and sources that I would never have found on my own. David was particularly useful because of his commitment to the project and his capacity to work without any clear direction from me.

I also owe a debt to Glenn Shean, Ph.D., professor of clinical psychology at The College of William and Mary. Glenn reviewed early drafts of key portions of the book and directed me toward a number of very useful contacts and resources.

The people at the Management Institute for Environment and Business (MEB) in Washington, D.C., helped me to clarify many questions in the early stages of my work. I am especially appreciative of the time that Patricia Choy and Christopher Cummings spent with me at their offices. Scott Fenn at the Investor Responsibility Research Center (IRRC) was kind enough to let me review the research of IRRC that related to the environmental performance of companies. My time with him greatly improved my own ability to understand what responsible environmental performance means. Andrew Mastrandonas, program director for the Global Environmental Management Institute, greatly expanded my understanding of the level of effort being made by organizations to improve their environmental performance.

Several colleagues read drafts of the manuscript and all of them gave me help: Lou Tagliaferri, Ph.D., president of Talico in Jacksonville Beach, Florida; John Brabson, Ph.D., professor of biochemistry at Mills College in Oakland, California; James Batchelor, Ph.D., chemical engineer, retired from the Department of Energy; Betty Batchelor, a careful critic and friend; Jack Clark, Ph.D., biologist; Steve Harris, assistant general manager of engineering, EG&G Florida; Charles Venuto, environmental manager, United Technologies (USBI); and Claire Kinlaw, Ph.D., molecular biologist and tree geneticist with the U.S.D.A. Forestry Service.

In writing this book, I made extensive use of the information data bases available at the Kennedy Space Center's Library and would like to thank Christal Wood, of the library's staff, for her technical assistance.

Stella Kinlaw, my wife, was her usual great help in reading and rereading the many drafts of the book. Finally, I would like to thank again Arlette C. Ballew, my editor at Pfeiffer & Company, for her efforts to turn one more manuscript of mine into a coherent and readable creation.

PREFACE

This book is a call to informed, thoughtful, and long-term action. It is a call to all managers and employees to begin to act in their own best interests by leading their organizations toward performance that is environmentally responsible and fully sustainable in relation to the earth's resources.

The thesis of the book is simple: Organizations must become environmentally responsible, or "green," to survive. Only by managing and working green, by making the environment an explicit part of every aspect of the organization's total operation, can the leaders of an organization expect to maintain its competitive position and ensure its survival.

It is no longer enough for organizations to demonstrate the continuous improvement of their services and products. They now are under pressure to demonstrate their capacity to deliver services and products that are friendly toward the natural environment. In the last ten years, the environmental issue has become a major organizational concern. We are moving beyond a time in which competitiveness can be achieved through total quality management and customer satisfaction. The degree to which these concepts will have continued utility will be determined by the degree to which they are redefined to include the environment. The organizations that are staking the clearest claims to a future are those that now see the natural environment as their most indispensible supplier and their most valued customer.

In this age of total environmental quality, organizations are responding with various degrees of seriousness.

At the low end of the scale, we find organizations that are largely reactive and operating on the basis of short-term planning. The more reactionary and defensive organizations waste a lot of time and resources launching well-funded lobbying operations to reduce or limit legal requirements for clean air, clean water, waste management, reduction of pollutants, and so on. These organizations practice a variety of avoidance strategies while trying to meet the minimum requirements of enforcement regulations. Some of these reactionary and defensive organizations have tried a variety of public relations strategies to give themselves "green" images while continuing to do business as usual. Some of them have undertaken deliberate campaigns to deceive the public.

At the high end of the scale are organizations that operate on the basis of foresight and long-term planning and aggressively search for ways to improve their environmental performance. Such organizations have accepted environmental responsibility as a condition for doing business and for remaining competitive. They are not going green by responding only to regulatory constraints and abiding by the law; they are going beyond what the law requires. They are designing new production processes that no longer emit chemicals that pollute the air and damage the ozone. They are changing the ways in which they package their products in order to reduce waste and the disposal costs of their customers. They are reclaiming and reusing their own wastes and secondary products through a variety of recycling programs for paper, spent chemicals and agents, plastics, water, and so on.

The fundamental difference between the reactive organizations that are doing as little as possible for the environment and the proactive ones that are doing as much as possible for the environment is that the latter have moved away from the "end of pipe" strategy for managing waste, pollutants, and toxic wastes and have embraced a "life-cycle" strategy for managing wastes and pollutants. The proactive organizations have recognized what the environmentalists have been saying for decades: Waste and pollution don't go away (there is no "away"); they just change form or position.

Some organizations have a history of being socially responsible and have taken the leadership in responding to the environmental challenge. Other organizations have recognized that their competitive positions now depend on how well they respond to the environmental challenge. Whatever the immediate causes are for responding, and whatever the degree and speed with which it might respond, every organization will respond.

Those that take the initiative to bring their total manufacturing and business operations into full compatibility with the needs of the earth's ecosystems will serve both their own self-interests and that of the planet and all its inhabitants. Those organizations that remain inactive and spend their resources trying to resist the forces that demand responsible environmental performance are committing themselves to a strategy that is untenable, and they will not survive in the age of the environment. In the age of the environment:

- "Dirty" manufacturers will not be competitive.
- Venture capital will be available only to environmentally safe projects.
- The green consumer will be the dominant consumer.
- Environmental advocacy groups will grow in numbers and influence.
- Environmental laws will increase in scope, and their enforcement will be more strict.

- International agreements that take the environment into account will determine the shape of enterprise.

One reason that organizations have been slow to recognize the environmental challenge as an opportunity to improve their performance, rather than as one that threatens their survival, is the idea that environmentalists and business leaders represent totally divergent needs and points of view. We have created a false polarity that pits working against living. Even worse, we have developed a totally unusable set of assumptions that suggest that we must stop doing business in order to save the environment or that we must continue to work at the expense of the environment.

The new background for planning organizational survival is biological survival. The two goals are rapidly coalescing.

Organizations are the conduits that make the resources of food, clothing, and shelter available to people. We have ample proof that these humanly created conduits can be made into systems that serve both our need to earn a living and our need to live.

The reality is that organizations are a major source of the environmental problems that we have, but they also are our major source for solutions to those problems. Organizations contain the fiscal, technological, and management resources that can be found no place else.

I suppose that every book is a form of therapy for its author. This book certainly has been a healthy experience for me. I am now considerably more optimistic about what is going to happen to us and our natural environment than I was when I began my research.

The preparation of this book has led me to discover hundreds of examples of very creative and successful responses to the environmental challenge. There is an enormous amount of thought and work going on around the world in organizations of every kind to make agriculture and all forms of production environmentally more sound. I also have discovered that there is developing a vast network of consortia, societies, and alliances that represent a new level of cooperation among business leaders, educators, politicians, and environmentalists, and which has no parallel in modern times. The international response to environmental issues will prove to be the most powerful stimulus yet in building community across national borders and across organizational boundaries.

I now am convinced that our world is developing an environmental conscience. The process of developing this conscience has been seriously under way at least since the 1950s and the publication of Rachel Carson's *Silent Spring*. Carson brought to public awareness ideas that were initially discussed on our streets, but which are now discussed in the boardrooms of organizations. More and more decision makers have been forced to acknowledge that we cannot use the environment as a cost-free resource.

This book identifies business and industry as the major resource for caring for the environment and for beginning to undo the multitude of ills and threats that we consumers have created for ourselves and our children. It describes what organizations are doing and what they must do. Most important of all, this book describes how organizations can take the leadership for healing the planet, while positioning themselves for sustainable performance in this new age of the environment.

INTRODUCTION

The primary question that now presses on every form of public and private enterprise is how to remain viable and still do business in a way that is friendly to the environment. How organizations respond to the many challenges posed by this question will determine in large measure their competitive positions and survival. In the last five years, we have moved from the era of total quality management (TQM) to the era of total quality environmental management (TQEM). At an ever-increasing pace, the meaning of "quality" of services and products that organizations produce is being redefined to mean quality treatment of the environment.

This does not mean that the movement toward TQEM has not been underway for some time. It is just that the level of pressure being exerted by forces such as environmentalists, laws, the marketplace, and international and national political coalitions has compelled the movement to shift into high gear.

Total quality management is a total way of managing. It implies achieving quality in everything the organization does. More and more organizations are realizing that they cannot achieve total quality if they dump poisonous effluent into the water supply or acid-generating chemicals into the air—any more than they can expect to achieve total quality if they do not treat their people properly and respond to the special needs of a multicultural work force. As Maurice Strong, Secretary General of the United Nations Conference on Environment and Development (1992 Earth Summit), remarked:

Efficient enterprises are at the head of the movement to sustainable development. Corporations that are on the leading edge of a new generation of opportunities created by the transition to sustainable development will be the most successful in terms of profits and the interests of their shareholders. Businesses that are defensive, fighting yesterday's battles, will fall by the wayside and will be caught in the backwash of the wave of the future [1].

In the pursuit of total quality, businesses and industries in the United States and around the world are noticeably becoming more responsible in terms of their environmental effects, or "going green." It is becoming more and more evident to corporate leaders that becoming environmentally responsible is the next step in total quality and is a step that is absolutely necessary in order to stay competitive and profitable.

SustainAbility, a global environmental consulting firm with home offices in London, began conducting its *Greenworld Survey* in 1990. In 1992, the firm reported that the consensus among businesses the world over was that "The environmental challenge will be one of the central issues of the 21st century." SustainAbility also reported that among such international conglomerates as AT&T, Amoco, British Gas, Exxon Chemicals, ABB Flakt, Rohm and Hass, and Shell International, there was a strong tendency to take a more proactive stance toward environmental issues and to be in front of the trends [5].

In a survey conducted in 1991 by Abt Associates, of Cambridge, Massachusetts, 75 percent of the Fortune 200 firms surveyed claimed that the environment was a central strategic issue. Compliance was viewed as the top environmental challenge, but pollution prevention, product development, marketing, and measuring environmental performance were focuses for new investment [3].

In another survey of corporations conducted in 1990 by the firm of Deloitte & Touche and the Stanford University Graduate School of Business, both located in California, 93 percent of the respondents identified the environment as either important or critical to overall corporate development [4].

The greening of business and industry is happening at an ever faster rate, and it is happening globally. The global characteristic of the movement toward more responsible environmental performance is apparent not only in what organizations are doing but also in the growth of international, environmental business networks. A partial list of such international networks includes the following:

- U.N. Center for Transnational Corporations
- Business Council for Sustainable Development
- International Chamber of Commerce
- International Environmental Bureau
- International Network for Environmental Management

Competitive & Green

The theme of this book is reflected in its title, *Competitive & Green*. This title implies that organizations must not only go green, but that they can actually maintain and improve their competitive positions by going green. Throughout this book, two main points will be emphasized:

1. The sooner that organizations regard the environmental challenge as a competitive opportunity, the more likely it is that they will survive and profit.
2. It is by emphasizing the environmental challenge as an opportunity for profit that we can best bring

under control the harm that we have been doing to the environment.

Business is not a separate issue *from* the environment. Business is the central issue *for* the environment. The ways in which we do business reflect what we believe and value. Business also is the most powerful, contemporary force that we have for shaping the course of human events.

Business transcends the boundaries and limits of nationalism. Also, business often exerts a dominant influence in political and social decisions. We can no more separate the issue of jobs from the issue of climate warming than we can separate the issue of family disintegration from that of poverty. As William Coors, Chairman of Adolph Coors Company, has remarked:

> Treating the environment and the economy as competing systems is like diverting food from a mother to feed her infant child. Each fulfills the other; neither can be sacrificed without sacrificing the larger whole [2].

Growing environmental awareness is just the latest chapter in a larger awareness that resulted in the demand for total quality. If there is an issue at all between the concerns of environmentalism to protect and renew the natural environment and the concerns of corporations for the bottom line, it is that the changes that are taking place appear to be occurring too slowly for the environmentalists and too fast for corporate leaders.

The Need for an Integrating and Organizing Concept

The movement by organizations toward a more environmentally sensitive way of doing business is clearly underway. More and more organizations are entering the green marketing arena, and more and more organizations are developing environmental policies and initiating programs to reduce, recycle, and reuse. A growing number of organizations are already taking advantage of the many competitive opportunities that the challenge of the environment offers.

We have enough examples from the more enterprising organizations to conclude that a proactive response to the environmental challenge can strengthen organizations and their competitive position by

- Avoiding the costs of fines, cleanups, and litigation.
- Reducing the amount of materials used.
- Reducing the amount and costs of energy.
- Reducing the costs of waste handling and disposal.
- Creating new distribution opportunities to new customers who are more environmentally sensitive.
- Maintaining market shares with old customers who want more environmentally friendly products.
- Creating new products and services for new market opportunities.
- Having greater credibility with banks and other financial institutions.
- Maintaining eligibility for less expensive insurance.
- Reducing risks of major environmental disasters.
- Developing and importing new technology.
- Improving the organization's public image.

Even though the advantages of going green are becoming more obvious, the movement toward an environmentally sensitive way of doing business still tends to be fragmented, uneven, and reactionary—primarily focused on fixing problems after they have occurred. Even organizations that have developed environmental policies or undertaken extensive environmental initiatives have not necessarily done so from a rationale that is sufficiently descriptive to communicate to the people in these organizations *why* they must work with the environment and *how* they must work with the environment.

Organizations need a way to gain the commitment of their people to new kinds of performance goals and new kinds of performance criteria that link quality, profit, and the environment together. Organizations need a concept or organizing principle that can assist them to plan, design, and undertake a thorough reordering of the ways in which they conduct every aspect of their business, starting with what they input from suppliers and ending with what they output into the environment.

Sustainable Performance

The concept that I propose as a vehicle for helping organizations to reshape themselves is called *sustainable performance*. I believe that this concept can serve as the following:

- A rationale that clearly describes why competitive performance and environmental performance cannot be in conflict.
- A framework for organizations to use in communicating to all their stakeholders (suppliers, employees, customers, and investors) how they intend to work for the environment, for profit, and for their own survival.
- An environmental planning and strategy guide.
- A tool for assessing and improving an organization's capacity to compete in the environmental age.

Sustainable performance is the central theme of this book and, along with the Sustainable-Performance Management Model (Figure 1-1), serves as its organizing principle. The greater part of this book is devoted to

1. Defining sustainable performance and its underlying principles.
2. Explaining why sustainable performance should become the all-encompassing goal toward which all organizations must strive.
3. Describing the management process for achieving sustainable performance.

For the moment, however, it is useful to develop a preliminary description of what sustainable performance means.

The term "sustainable performance" describes how organizations must conduct their business in order to continue conducting that business into the indefinite future. If the primary goal of organizations is to stay alive, then sustainable performance describes the requirements for staying alive in the new age of the environment. If quality performance has become the watchword of this decade, sustainable performance (or something similar) will become the watchword of the next.

Sustainable performance is both the goal toward which organizations must strive and the way in which they must plan, execute, and evaluate every aspect of their business. The Principles of Sustainable Performance must be evident in organizational planning at the highest levels as well as being evident in the way employees undertake their day-to-day work. Sustainable performance must be explicit in an organization's policies and strategic plans as well as being explicit in all its human resource systems, such as performance appraisals, selection and promotion, and reward systems.

Sustainable performance is the next evolution in the structure and functioning of organizations. It encompasses the unalterable requirement that organizations must accept as a condition of doing business. They must learn how to function as partners with the environment.

> Sustainable performance is the evolution of organizations into wealth-producing systems that are fully compatible with the natural ecosystems that generate and preserve life.

This concept of sustainable performance has been anticipated by such documents as the *CERES Principles,* issued by the Coalition for Environmentally Responsible Economies, and the *Business Charter for Sustainable Development* of the International Chamber of Commerce. The idea was cultivated by John Elkington and his colleagues at SustainAbility, Ltd., in London; and by the work of the United Nations Environment Programme; the International Institute for Sustainable Development; the Business Council for Sustainable Development; and research groups such as The Center for Environmental Management at Tufts University in Massachusetts.

My contribution in this book is not so much to present a new concept as to integrate already existing ideas about sustainability and to translate the general concept of sustainable development into the more concrete idea of sustainable performance. I have also provided practical tools and guidance to help organizations move toward sustainable performance.

Purposes of the Book

This book describes why leaders must move their organizations toward sustainable performance and how they can do it. It is for anyone who wants to take the lead in helping his or her organization respond to the environmental challenge. It provides guidance and tools to assist leaders at every level to undertake the practical, day-to-day business of sustainable performance.

I also intend this book to be a resource for educators and consultants who have the responsibility to equip employees and other stakeholders of organizations with the knowledge and tools they need to participate actively in planning and undertaking sustainable performance.

We are in the midst of a new phase of the worldwide quality movement that will change forever the way we do business. In this new era, the environment will take precedence as industry's most important supplier and most valued customer. Nature will become the final arbiter for determining what is and what is not quality.

In the move from total quality management to total quality environmental management, every person at every level of responsibility must learn the new rules of green management and become involved in the struggle toward sustainable performance.

I believe that this book is a tool that can create such involvement.

General Description and Overview of Content

In Chapter 1, I provide an overview of the Sustainable-Performance Management Model and outline the process for building sustainable performance into an organization. The model becomes the framework for understanding the information that is presented in the following chapters.

Chapter 2 describes the specific milestones that can be used to plan and manage the process of achieving sustainable performance.

Chapter 3 describes the many pressures that are forcing organizations to choose between going competitive and green or just going away.

Chapter 4 describes the Systems Model for Sustainable Performance and the characteristics of sustainable performance.

Chapter 5 contains a set of values associated with sustainable performance and summarizes these values in the Principles of Sustainable Performance.

In Chapter 6, I introduce the Response Model for Sustainable Performance and discuss the competencies and actions required for managing an organization's response to the challenge of sustainable performance.

Chapter 7 discusses the general strategies and special tools of sustainable performance.

Chapter 8 reviews the Sustainable-Performance Management Model and summarizes the key leadership and management tasks and competencies required for taking responsibility for sustainable performance.

The Appendix contains copies of the *CERES Principles* and the *Business Charter for Sustainable Development* of the International Chamber of Commerce. The Appendix also contains a summary of the more important environmental laws, a copy of the *Sustainable-Performance Assessment*, a list of special resources for sustainable performance, and other useful information.

Although this book is largely based on the changes that are taking place in the environmental performance of organizations in the U.S., it does have a larger reference. The idea of sustainable performance, which forms the central theme of the book, has universal application. Where appropriate, I've included international examples to argue my case for and to illustrate the concept of sustainable performance.

References

1. *Business and the Environment.* (June, 1992). p. 4.
2. Coors, W.R. (n.d.). *The Myth That Divides Environmentalists and Industry.* Golden, CO: Adolph Coors Company.
3. Corporate Environmentalism. (December 1991). *Environmental Business Journal,* p. 1.
4. *The Economist.* (September 8, 1990). p. 18.
5. Elkington, J. (1991). *The Corporate Environmentalists: A Report on the 1991 Greenworld Survey.* London: SustainAbility Limited.

1

THE SUSTAINABLE-PERFORMANCE MANAGEMENT MODEL

The purpose of this chapter is to develop an overview of the knowledge, skills, and steps required of people who want to help their organizations move toward sustainable performance. I will develop this overview by presenting the Sustainable-Performance Management Model (Figure 1-1). Once the model is understood, it can be used to integrate the rest of the information presented in this book and as a practical tool for planning and implementing sustainable performance.

The model provides the structure for the way in which the material in this book is organized. After the general description of the model in this chapter, each subsequent chapter discusses in detail the various elements in the model.

Before introducing the Sustainable-Performance Management Model, it is important to make the idea of sustainable performance as concrete as possible. One way to do this is to review a few examples of what organizations are actually doing to perform more responsibly toward the environment and, at the same time, improve their own competitive positions and profitability.

Examples of Sustainable Performance

No organization currently can provide us with a complete demonstration of sustainable performance. There are, however, a large number of examples in the United States and elsewhere that illustrate how organizations are improving their environmental performance and their bottom lines.

Saving energy by conservation and increased efficiency in the processes of production is one of the most obvious opportunities for improved environmental performance. The capacity to reduce the use of energy for industrial purposes has been well demonstrated. Between 1973 and 1990, the U.S. economy grew about 40 percent, while energy consumption grew by only 5 percent—saving about $160 billion in energy costs. China has achieved a 3.7 percent reduction of industrial energy use during the 1980s and is expected to continue a similar rate in the 1990s [10].

El-Nasr Glass and Crystal Company is the largest Egyptian glass-producing company. An analysis of the energy use of one of its major furnaces revealed a 50 percent loss of energy during combustion. Modifications to the furnace lowered energy requirements by 55 percent, increased furnace productivity by 255 percent, and resulted in an annual savings of approximately $800,000 [11].

Organizations are finding enormous opportunities for saving money in the ways in which they package their products. Packaging uses energy and materials in its production, increases the costs of transporting products, and produces waste at the customer end. Organizations are finding that they can reduce costs for themselves and their customers by packaging their goods in ways that are more environmentally friendly.

The Hewlett-Packard Company (HP) has changed its shipping cartons from bleached white boxes to unbleached kraft boxes as part of its program to encourage company-wide environmental practices. The switch was employee inspired and began as a grassroots effort in several HP divisions. The kraft boxes contain a higher percentage of recycled material, are less expensive, and are less likely to be discarded because of scuff marks on their surfaces. HP estimates that the switch will save the company approximately $3 million per year.

HP has made additional changes in its packaging. It estimates another savings of $2 million per year from switching to a less toxic ink on all its containers and from reducing the number of boxes that it uses to ship some of its equipment. To demonstrate its full commitment to the environment, HP's New Jersey power-supply division pays for its customers to send its shipping cartons and foam cushions to recyclers [3].

Digital Equipment has developed a "packaging waste management program" that will eliminate 5,400 tons of packaging from its operations and cut $30 million from the cost of handling it.

One proven strategy for saving money is recycling. The hotel industry, for example, is finding ways not only to recycle wastes but also to recycle resources such as water. Hyatt Hotel and Resorts estimates that it will save $3 million each year through its recycling program and will reduce the amount of garbage sent to landfills by 30 percent. A hotel in Hawaii recycles water from its laundry into the air-conditioning system. The garbage-collection bill for the Hyatt Regency Chicago in 1989 was $12,000 per month. As a result of the hotel's recycling program, the bill is now $2,000 per month.

The Rexham Corporation provides an example of how chemicals and other materials can be reclaimed and reused. The company installed a $16,000 distillation unit to reclaim n-propyl alcohol from waste solvent. The unit recovers 85 percent of the solvent in the waste stream, which results in an annual savings of $15,000 in virgin-solvent costs and $22,800 in waste-disposal costs.

Organizations often generate toxic wastes as side effects in their production processes. Paying for the storage and disposal of these wastes is becoming more and more expensive. Not having to pay for the handling of any kind of waste is to an organization's advantage. Managing waste and the costs related to it are non-value-added activities.

DuPont now spends about $1 billion per year on recycling and reducing wastes. In 1991, these efforts helped to generate more than $100 million in revenue and cost savings combined. DuPont claims that the figure will rise to $1 billion per year by the end of the decade. [12].

Chevron introduced a toxics-reduction program in 1987 and encouraged employees to suggest ideas to eliminate waste. By 1991, the company had reduced hazardous waste by 60 percent (compared to 1986 levels) and saved $10 billion in handling and disposal costs [13].

Northern Telecom, Inc., in North Carolina, eliminated chlorofluorocarbons (CFCs) from its manufacturing process in January, 1991, and set a goal to eliminate them in all its plants within a year. CFCs had been used by the company in its manufacture of telecommunications equipment, to clean circuit boards after they had been soldered. The CFCs removed leftover flux. To end CFC use, Northern Telecom switched to an alcohol-based flux that does not require cleaning because its residue evaporates during production. Developing the new process cost the company about $1 million, but eliminating CFCs already has saved more than $4 million in cleaning costs. Northern Telecom expects to save over $50 million in the next eight years.

Reducing solid and hazardous wastes by changing packaging and eliminating the use of chemicals that have harmful environmental impact are examples of performance that are profitable and environmentally responsible, i.e., sustainable. Another example is the reduction or elimination of pollutants such as carbon dioxide, the primary "greenhouse" gas.

Geneva Steel, based in Utah, is one of the few steel makers in the United States that is profitable. It ranks among the most profitable steel makers in the world. Geneva Steel also has a good environmental record. The company has cut emissions from its coke ovens to less than a quarter of the permitted maximum while at the same time improving the efficiency of its furnaces and reducing the costs of production [9].

Throughout the power-production industry, the move is on to reduce the use of energy and the emission of carbon dioxide. However, it is not a newfound social and environmental conscience that has prompted the industry to become environmentally responsible. The impetus is to stay in business and turn a profit for investors. The traditional way in which power utilities have made money is by selling more power. Under current rules, however, they often can make as much money from promoting conservation as they can from building new power plants. In the state of California, for example, power companies that cut costs by generating less power are permitted to pass 85 percent of the savings on to their customers and keep 15 percent for themselves.

The fundamental characteristic of sustainable performance is that it *sustains the environment* and *sustains the organization's bottom line.* There is enough evidence now to prove that sustainable performance is an achievable goal and that many organizations are well on their way to achieving that goal. The U.S. Environmental Protection Agency has on file thousands of examples of companies that are making or saving money by improved environmental performance [6].

The preceding examples focus on making money by avoiding pollutants and waste and by reducing the costs of managing them. As organizations progress toward sustainable performance, however, they will become more involved in new technologies such as designing-for-environment (DFE) and closed-loop-processing (CLP). DFE is the developing technology of designing all-new facilities and processes that have the most limited negative impact possible on ecological systems and that are aesthetically integrated into local environments. In the building industry, DFE sometimes is called "green architecture." CLP is the technology of retrofitting old production and manufacturing processes and building new ones that use the waste from one operation to fuel another operation down the line.

Baxi Heating is a manufacturer of domestic boilers in the United Kingdom. Baxi built a new foundry in the middle of a small town in Lancashire, England, that overcame the traditional problems of aesthetics, fumes, smell, and noise. This highly innovative example of designing-for-environment not only produced a plant that was compatible with the needs of people and the environment but also saved the company $4.5 million by allowing it to build where roads and services already existed rather than having to build in isolated areas and paying to install its own roads and utility lines. Also, to avoid problems in waste management, the company designed a waste-recovery system to reclaim clays and minerals from its waste sludge. This system now saves Baxi around $125 million per year in reclaimed materials and waste disposal [6].

The examples of sustainable performance presented here provide only a suggestion of what organizations all over the world are doing to improve their environmental performance and their competitive positions. However, although there are an enormous number of such examples, the change to sustainable performance still is often fragmentary, uneven, and unorganized. What is needed is a tool that can help people to grasp the total process involved in the movement of organizations toward sustainable performance. The Sustainable-Performance Management Model (Figure 1-1) is such a tool.

The Sustainable-Performance Management Model

Managing for sustainable performance is a process that begins with understanding and ends with improved performance. It is imperative that leaders take the time to understand the full scope of this process before they begin to make decisions and take actions to help their organizations respond to the environmental challenge. The Sustainable-Performance Management Model (Figure 1-1) can provide such understanding.

Each of the elements in the model will be identified and described in this chapter. Then each element in the model will be fully explained in later chapters of the book. Each of the "milestones for sustainable performance" will be discussed in Chapter 2. The "pressures" on organizations to force them to perform responsibly toward the environment are discussed in Chapter 3. The "characteristics of sustainable performance" and the "Systems

20 / Competitive & Green

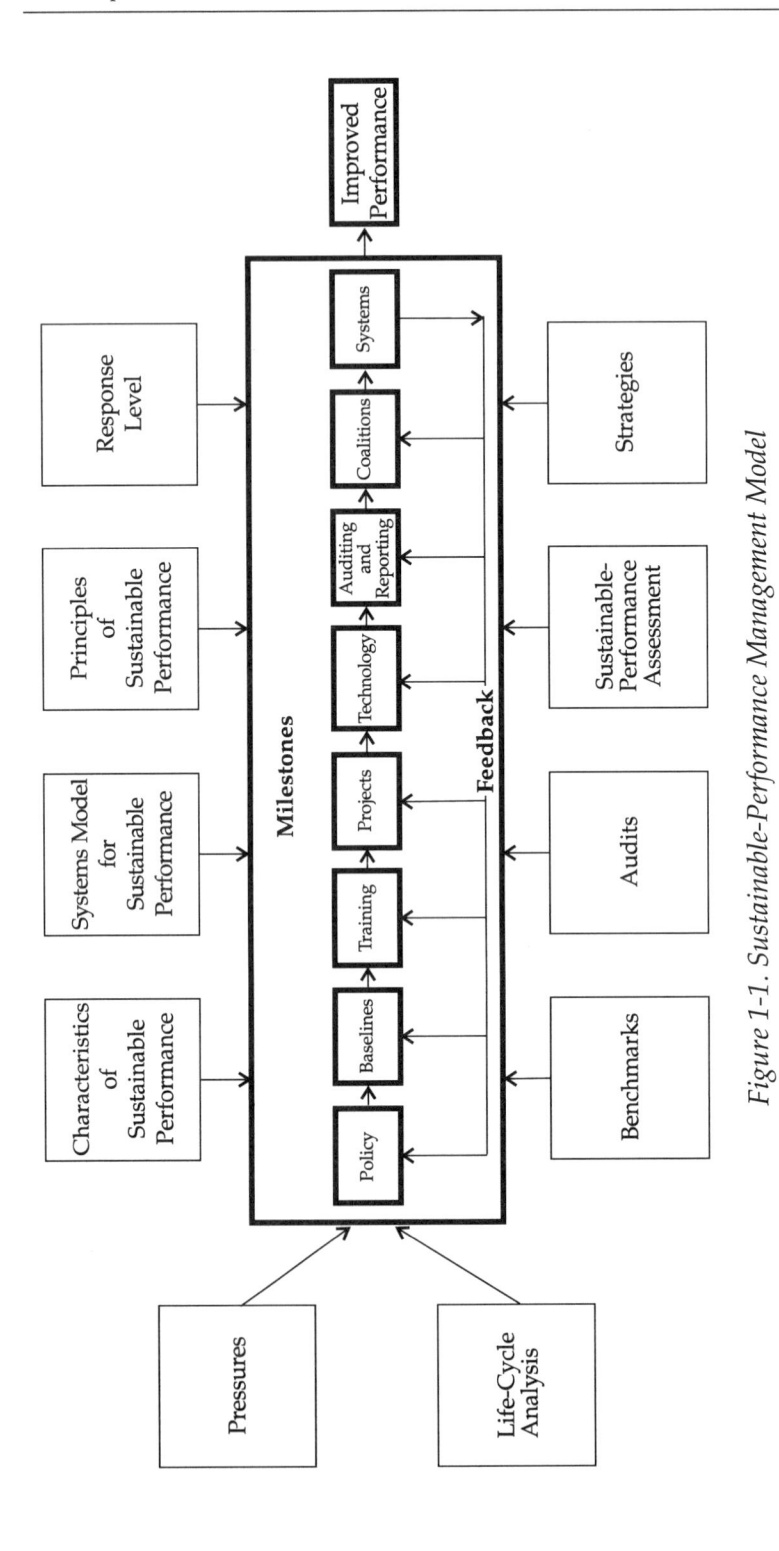

Figure 1-1. Sustainable-Performance Management Model

Model for Sustainable Performance" are discussed in Chapter 4.

The model will be referred to throughout the book and will be used to summarize the discussion of sustainable performance at the end of the book. The purpose of presenting it here is to provide the reader with a framework with which to organize the information that is contained in the rest of the book. My goal is to help the reader to never lose sight of the big picture. The book is about the process of managing for sustainable performance. It is necessary, however, to cover a number of different topics and to develop a large amount of information. The model will help to organize these topics and information into a connected body of practical management knowledge.

Uses of the Model

The Sustainable-Performance Management Model has the following specific uses:

- It provides a graphic display of the various events that must take place from the point at which sustainable performance is recognized as a goal to the point of improved performance.
- It identifies the key elements in the process of building sustainable performance and shows how these elements are related.
- It helps leaders to take a systemic and holistic view of sustainable performance and to avoid focusing on single events or actions.
- It conveys the fundamental idea (through the feedback loop) that sustainable performance is a process of continuous improvement.

- It gives leaders a concrete basis for discussing sustainable performance and communicating their intentions to their organizations.
- It keeps leaders aware of the skills and knowledge that they need to manage sustainable performance.

Description of the Model

The model has a central flow that consists of the major *milestones* that must be reached as an organization moves toward sustainable performance. This flow begins with the formulation of an environmental *policy* and ends with building the necessary management and human resource *systems* required to support sustainable performance. The achievement of the milestones results in *improved performance*. These milestones are shown as sequential events for the purpose of providing a simple description of the flow. In actual practice, many of the milestones will be worked on simultaneously.

All the milestones are linked together by *feedback* on the quality of the improved performance achieved. As results are monitored and evaluated, milestones are revised.

The milestones are surrounded by a set of *sources* for the knowledge and skills required to achieve the milestones and manage sustainable performance. Reading clockwise in the model, these are pressures, characteristics of sustainable performance, Systems Model for Sustainable Performance, Principles of Sustainable Performance, response level, strategies, Sustainable-Performance Assessment, audits, benchmarks, and life-cycle analysis. The model suggests that all these sources provide input to the total sequence of initiating sustainable performance from *policy formation* to *improved performance*.

The Milestones

The use of milestones conveys the idea that the initial movement toward sustainable performance can be thought of as a project and managed as a project. Thinking of the process of moving toward sustainable performance focuses the organization on the achievement of specific goals and helps it to allocate the necessary resources to achieve those goals.

Milestones denote major developmental targets within projects as well as showing the time when such targets will be reached. Milestones are proven tools for managing a goal-related process.

The milestones used here have been derived from a careful analysis of what companies that are leaders in improving their environmental performance have done and from the work of various national and international coalitions that are serving as clearing houses and resources on environmental performance.

The milestones for sustainable performance are discussed at length in Chapter 2. They are

1. **Milestone One:** Sustainable-performance policy statement is published.
2. **Milestone Two:** Sustainable-performance baselines are established.
3. **Milestone Three:** Initial sustainable-performance training is accomplished.
4. **Milestone Four:** Initial improvement projects are underway.
5. **Milestone Five:** Development of environmental technologies are being supported.
6. **Milestone Six:** Auditing and reporting system is functioning.

7. **Milestone Seven:** Coalitions exist.
8. **Milestone Eight:** Management and human resource systems are revised to support sustainable performance.

The Sustainable-Performance Management Model also depicts a set of information sources that are required for developing the competencies in leaders to manage the flow through these milestones. All these information sources are described in later sections of the book. They are explained briefly in the following sections in order to provide a complete overview of the model.

Pressures

Many social and economic pressures are forcing organizations to respond to the environmental challenge. To develop an appropriate environmental policy and to take responsibility for responding to these pressures, leaders first must understand them. These pressures are discussed at length in Chapter 3. They are

1. **Compliance.** The increasing number and stringency of laws and regulatory requirements.
2. **Punitive Fines and Costs.** The fines for noncompliance and the costs associated with responses to accidents and disasters are growing in frequency and amount.
3. **Personal Culpability and Imprisonment.** Individuals are being fined and threatened with imprisonment for violating environmental laws, and more and more such laws are being passed or are on the books.
4. **Environmental Activist Organizations.** There has been a proliferation of such groups and their reform

agendas at the international, national, state, and local levels.

5. **An Aroused Citizenry.** Citizens are becoming informed through the media and more substantive sources and are seeking a variety of channels through which to voice their desires to business and industry.

6. **Societies, Coalitions, and Associations.** Professional societies, trade associations, and various *ad hoc* coalitions are making pronouncements and initiating programs that influence organizational behavior toward the environment.

7. **International Codes for Environmental Performance.** The *Valdez Principles* issued by The Coalition for Environmentally Responsible Economies and the *Business Charter for Sustainable Development* developed by the International Chamber of Commerce are creating global pressures for environmentally responsible performance.

8. **Environmentally Conscious Investors.** Organizations are recognizing that their environmental performance and their potential financial risks for poor performance (fines, cleanup costs, litigation expenses) will determine how attractive their stocks are to investors.

9. **Consumer Preference.** Consumers are looking for green practices and green products, and organizations must respond with more than green hype and advertising.

10. **Global Markets.** International competition exists within the context of a multitude of environmental laws that will no longer permit organizations from

developed countries to export their pollution to developing countries.

11. **Global Politics and International Organizations.** A variety of international organizations and forums are exerting direct pressure on nations, which invariably affects business and industry.
12. **Competition.** The pressure that stands at the intersection of all others is from those companies that are embracing sustainable performance and improving their competitive positions.
13. **Other Pressures.** First, qualified people increasingly will prefer to work in organizations with good environmental records. Second, in the future "full cost pricing" will require that companies reflect in the prices of products and services not only production and delivery costs but the full costs of environmental degradation associated with the products and services.

To manage the process of sustainable performance, leaders must understand the many pressures that are forcing organizations to respond to the environmental challenge. They must have a full and pragmatic understanding of the meaning of sustainable performance. This book provides the following three sources of information about sustainable performance:

- The Characteristics of Sustainable Performance
- The Systems Model for Sustainable Performance
- The Principles of Sustainable Performance

Characteristics of Sustainable Performance

Knowing the full meaning of sustainable performance is a critical input for putting together an environmental policy and a fundamental requirement for the competent management of the total process. In Chapter 4, I describe how sustainable performance is similar to sustainable development (a more general term that has become extremely popular) and explain the distinguishing characteristics of sustainable performance. These two characteristics are profit and performance.

Systems Model for Sustainable Performance

The traditional open-systems model of organizations includes three elements: input, transformations (work processes), and output. More expanded models may include suppliers and customers. None of the traditional models, however, include the environment. The environment now must be taken into account; along with the traditional elements in an open-systems model, it provides the basis for understanding sustainable performance.

The Systems Model for Sustainable Performance is presented in Chapter 4 (Figure 4-1). This model is a primary tool for understanding just how different sustainable performance is from traditional performance. The model shows the environment as the context within which organizations must plan and execute performance. It shows the environment interacting with the three traditional elements of input, process, and output. It also suggests that the environment creates a new dimension for the ways in which suppliers and customers are perceived and managed.

To manage sustainable performance, leaders must learn a new set of competencies. The model suggests that the pressures to change, the characteristics of sustainable performance, and the Systems Model for Sustainable Performance are sources of the information from which such competencies can be gained. Another source is the Principles of Sustainable Performance.

Principles of Sustainable Performance

Sustainable performance, like total quality management (TQM), represents a major change in the underlying values or principles of doing business. To understand the nature of this change and to formulate a useful company policy, leaders must understand this change in values. The Principles of Sustainable Performance summarize these values and turn them into a set of guidelines that can help leaders to manage the process of sustainable performance. These principles are explained fully in Chapter 5. They are as follows:

Principle One. Sustainable performance is a process of systems thinking, analysis, and integration that requires that the organization be understood and managed as a system.

Principle Two. Sustainable performance is an ecologically interdependent process and requires that all organizational processes, products, and services be revised or replaced to ensure their compatibility with nature's ecosystems [6].

Principle Three. Sustainable performance is a results-oriented process and requires the demonstrated commitment of organizational leaders to specific, measurable results.

Principle Four. Sustainable performance is a community-building process. This requires organizations to cooperate with one another and use the environment in ways that are equitable for one another. This also requires that organizations involve all their stakeholders in the processes of planning and implementing sustainable performance.

Principle Five. Sustainable performance is a limiting process. It requires that organizations recognize that there are costs associated with the earth's resources and ecosystems that must be included in the organizations' accounting processes and which will place limits on the size and nature of their businesses.

Principle Six. Sustainable performance is an open process and requires that organizations communicate fully all aspects of their planned and actual environmental performance to all the organizations' stakeholders [1].

Principle Seven. Sustainable performance is a process of continuous improvement of every aspect of an organization's performance and requires the full involvement of every member of the work force [4,5].

Principle Eight. Sustainable performance is a data-based process and requires concrete information retrieved from auditing, measuring, and reporting the organization's environmental performance [1].

Principle Nine. Sustainable performance is a technologically dependent process and requires organizations to develop partnerships with governments, other organizations, educational entities, research and development sources, suppliers, and customers in order to discover and implement ways to improve sustainable performance [2,3].

Principle Ten. Sustainable performance is a total organizational process and requires that all planning, decision making, and human resource systems be made fully congruent with the organization's commitment to sustainable performance.

In addition to knowing what the pressures are and understanding the meaning and content of sustainable performance, leaders need to understand how their organization is responding to the environmental challenge and what the organization's potential for response is.

Response Level

The response level that an organization achieves is a function of a number of variables such as the *screens* that leaders employ in the ways in which they access and use information about the environment; the *competencies* that people in the organization have to undertake sustainable performance; and the *motives* that are dominant in an organization for responding to the environmental challenge. The levels that typify an organization's response are

1. Compliance with the law.
2. Nonintegrated initiatives.
3. Integrated environmental plan and initiatives.
4. Sustainable performance.

These response levels are discussed in Chapter 6, along with the competencies that leaders need in order to manage their organization's total response to the environmental challenge.

The first concern of leaders who want to manage the process of sustainable performance is to help their organization to formulate and publish a policy that communicates exactly what the organization believes about its environmental responsibility and what it intends to do to improve its environmental performance. A related concern is to understand the milestones that must be reached and to plan the development of sustainable performance around these milestones.

Strategies

Managing for sustainable performance requires an understanding of the strategies and tools that are useful in improving environmental performance. These strategies and tools are described in detail in Chapter 7.

Tools

- The Sustainable-Performance Assessment (SPA)
- Audits
- Benchmarks
- Life-Cycle Analysis

Strategies

1. Practicing conservation and paying attention to every detail associated with a work process. This means using only the necessary amounts of materials, turning off the water and lights, keeping machinery and vehicles in top running condition, and so on.

2. Modifying or replacing existing processes, products, and services to make them environmentally friendly. Common examples are changing to more energy-efficient machinery, reducing packaging materials, and eliminating toxic chemicals and emissions.

3. Reclaiming by recycling and reusing waste and secondary products such as chemicals, paper, plastic, metal, and water.

4. Reducing the use of materials, for example, reducing the amount of packaging or packing used,

reducing the size of reports and invoices, reducing the amount of materials used in a process, and reducing energy used.

5. Finding new "green" niches in the marketplace and delivering new services and products, such as waste management and disposal, alternative energy sources, alternatives to toxic and ozone-depleting chemicals, and benign cleaning agents.

The Sustainable-Performance Assessment

The Sustainable-Performance Assessment (SPA) is based on the Principles of Sustainable Performance and the milestones for sustainable performance. The SPA is a tool that can help an organization to track its performance toward sustainable performance. When used over time, the SPA provides an index of performance and helps the organization to focus on the entire process of building sustainable performance into the organization. Data from the SPA leads naturally to the formulation of improvement targets and the development of specific improvement projects.

The SPA can be used in the following ways:

1. **As an organizational assessment tool.** In this case, the whole work force (or an appropriate sample) completes the SPA, and the data is used to develop a baseline (basic standard or level) of the organization's response.

2. **As a management-feedback tool.** In this case, managers use the SPA to communicate their own perceptions of how the organization is responding to the environmental challenge.

3. **As a third-party audit tool.** In this case, the audit team uses the SPA and from interviews, analysis of documents, and direct observation provides an

assessment of the organization's level of environmental performance.

Audits

Baselines are basic standards or levels by which to judge future performance. To establish baselines, one must know what the measured environmental performance of an organization is. The most systematic way to do this is through audits. Audits—as distinguished from risk assessments—are tools for involving the whole work force in the ongoing process of continuous improvement. Audits and the process of auditing are explained in Chapter 7.

Benchmarks

Benchmarks are "known points of elevation." As used to identify improvement opportunities, benchmarks represent the best known performance in a specific area, such as engineering design, toxic-emission control, or reduction of packaging waste. Benchmarks and the process of benchmarking are discussed in the section on tools in Chapter 7.

Life-Cycle Analysis

Life-cycle analysis (LCA) is a process of analyzing all of the inputs and outputs to determine the total environmental impact of the production and use of a product. LCA utilizes a set of inventories that quantifies the use of energy, the use of resources, and environmental releases into the air, water, and land. More complete information about LCA is found in Chapter 7.

Additional Tools

The Sustainable-Performance Management Model does not specifically include the many tools that are associated with total quality management. Sustainable performance builds on the commitment to total quality. I assume that any organization that intends to embrace sustainable performance and total quality environmental management is also equipped for TQM. These tools include

- **Models.** These are pictures or other representations of procedures or processes that assist teams in understanding and planning improvement.
- **Rational/Structured Tools.** These are tools that provide an orderly process for accomplishing tasks such as developing information, identifying likely causes, creating alternatives, and making decisions. Examples are brainstorming, flow charting, interviews, and cause-and-effect diagrams.
- **Numerical/Statistical Tools.** These are tools that can be used to count, to show mathematical relationships, and to measure the performance of work processes. Examples are run charts, Pareto charts, histograms, surveys that produce data that can be statistically manipulated, and control charts.

The Sustainable-Performance Assessment, audits, benchmarks, and life-cycle analysis are shown as special sources of information in the Sustainable-Performance Management Model. This is to emphasize them as special sources of information throughout the entire process of sustainable performance.

Summary

The Sustainable-Performance Management Model will be referred to throughout this book. It should be consulted often as the reader progresses through the information contained in the following chapters. Everything that follows in this book is an explication and expansion of the model.

The model describes a process that flows through a set of milestones. This flow is informed by information about the following:

- Pressures
- Characteristics of sustainable performance
- Systems Model for Sustainable Performance
- Principles of Sustainable Performance
- Response level
- Milestones for sustainable performance
- Strategies
- Sustainable-Performance Assessment
- Audits
- Benchmarks
- Life-Cycle Analysis

The model will be used to develop three parallel sets of information:

1. An understanding of a set of *milestones* that can be used in managing the process of sustainable performance.
2. An understanding of *sources* of information that can provide leaders with the knowledge and skills for managing sustainable performance.
3. An understanding of key *competencies* that leaders need to manage sustainable performance.

In the next chapter, I will discuss the milestones for sustainable performance. In subsequent chapters, I will describe the sources of information and the competencies that these sources can help leaders to develop for managing sustainable performance.

References

1. *Audubon.* (July, 1990). pp. 108-11.
2. Business and the Environment. (November 8, 1991). p. 5.
3. Business and the Environment. (December 20, 1991). p. 3.
4. Coors, W.R. (n.d.). *The Myth That Divides Environmentalists and Industry.* Golden, Colorado: Adolph Coors Company.
5. Corporate Environmentalism. (December, 1991). *Environmental Business Journal,* p. 1.
6. Elkington, J., & Knight, P., with Hailes, J. (1991). *The Green Business Guide.* London: Victor Gollancz, Ltd.
7. The Economist. (September 8, 1990). p. 9.
8. The Economist. (September 8, 1990). p. 18.
9. Barriers Facing the Achievement of Ecologically Sustainable Industrial Performance. (1991). *Proceedings of the Conference on Ecologically Sustainable Industrial Development.* Vienna, Austria: United Nations Industrial Development Organization.
10. Scoullos, M. (January-March, 1990). Energy Conservation and Efficiency: A Safe Response to Global Warming and Other Environmental Problems. *Industry and Environment, 13*(1), 6-8.
11. Selim, M.H. (April-June, 1990). Promoting Energy Efficiency in Egyptian Industry. *Energy and Environment, 13*(2), 15-18.
12. *USA Today.* (May 15, 1991). p. B3.
13. World Resources Institute. (1991). *World Resources 1991.* Washington, D.C.: Author.

2

MILESTONES FOR SUSTAINABLE PERFORMANCE

The Sustainable-Performance Management Model (Figure 1-1) describes a set of milestones that lead to sustainable performance (SP). This chapter will focus on those milestones. It also will introduce the competencies that are required to manage SP. These competencies for leaders include the following:

1. Leading their organizations through the milestones that must be reached in their progress toward SP.
2. Possessing a clear understanding of the pressures forcing organizations to move toward SP.
3. Having a clear understanding of the meaning of SP, its characteristics and principles.
4. Understanding the level of responses that organizations can be expected to make to the environmental challenge.
5. Knowing what level of response their own organization is making.
6. Being able to influence the level of response their organization is making and lead it toward SP.
7. Understanding and using the general strategies and special tools for SP.

In this chapter, I will describe the first competency and the milestones that are intermediate events or targets along the path to SP.

The kind of change that organizations should strive to achieve is to make SP so fully integrated into the day-to-day operations of the organization that it is no longer perceived as something special but is the normal way of doing business. As with any major transformation, it is useful, at least initially, to think of the transformation as a project and to manage it as one would manage a project.

Derivation of Milestones

The use of milestones has been popularized by the increased use of projects. A project is a way of focusing an organization on the achievement of specific goals and of allocating the necessary resources to achieve these goals. Projects are used to develop or modify facilities, products, and systems and to achieve a host of other results.

Milestones denote major developmental targets within projects and show when such targets will be reached. Milestones also convey information about the sequence of key targets and their relationship. Finally, milestones express how far a project has progressed and how far it is from completion.

The SP milestones are derived from an analysis of a wide range of sources from professional and expert groups such as the following:

- Coalition for Environmentally Responsible Economies
- Council for Economic Priorities
- Environmental Defense Fund
- Global Environmental Management Initiative
- International Chamber of Commerce

- Management Institute for Environment and Business
- Natural Resources Defense Council
- SustainAbility, Ltd.
- United Nations Industrial Development Organization
- World Commission on Environment and Development

In addition, I have analyzed the environmental performance of numerous companies that have been mentioned in various publications as having established themselves as leaders in the movement toward SP. These companies include: Anheuser-Busch, AT&T, The Body Shop, DuPont, Hewlett/Packard, IBM, Northern Telecom, and Proctor & Gamble.

Qualitative Characteristics of Management

Certain qualitative characteristics are evident in every management action and decision in managing an organization's movement through the milestones toward SP. These qualities are

1. Demonstrated commitment of management to sustainable performance.
2. Determination to involve the work force.
3. Determination to involve the organization's other stakeholders.

The Commitment of Management

The demonstrated commitment to SP of senior executives and managers at all levels of an organization is required if the movement toward SP is to succeed in the organization.

Commitment to SP carries the same kind of imperative that commitment to TQM and continuous improvement requires. The change to TQM and the change to SP both represent such radically new ways of doing business that they require the specific and visible attention of managers at all levels. The degree of change demanded for SP will be emphasized further in the discussion of the Principles of Sustainable Performance in Chapter 5.

TQM and SP are survival strategies. Both represent radical revisions in an organization. However, executives and managers have sometimes failed to demonstrate the needed level of commitment to institutionalize TQM for the same reasons that they may fail to demonstrate the level of commitment required to institutionalize SP. The basic problem is that they fail to grasp the radical nature of the change that is required.

Failure in demonstrating commitment occurs because executives and managers see SP, like TQM, as something that can fit comfortably in current organizational structures and be accommodated by old organizational values without any changes except a few labels, titles, and a new staff member or two.

Sustainable performance requires changes in every aspect of the organization's business. It will change the kinds of strategic and business plans that are made, the ways in which resources are allocated, the rules of competition, and the equation for profit.

Commitment by executives and managers has two major impacts on an organization. It communicates what is important, and it tells the work force where people should focus their energies. Executives and managers communicate their commitment through the following behaviors:

1. They demonstrate willingness to learn the new competencies that are required by the new way of

doing business. They lead the way in gaining new skills and in coaching others about these new competencies. They attend conferences on the environment, and they sponsor learning events on sustainable performance. Sustainable performance will make slow progress in an organization if executives and managers do not demonstrate their willingness to learn the kinds of competencies discussed throughout this book.

2. They engage in rituals such as media interviews, publications, speeches, and awards. Every time organizational officials make speeches, give interviews, preside at company celebrations, or introduce a management-education program or an executive retreat, they communicate what is important to them and what should be important to other people. They will communicate what is important to them, whether they like it or not, by what they say and do and what they do not say and do. It goes with the job.

3. They demonstrate consistency and perseverance. It is what executives and managers are perceived to be doing on a routine basis, day in and day out, that ultimately will communicate how serious they are about SP.

The following are a few ways in which commitment to SP can be communicated by executives and managers:

1. Ensure that the environment is a priority at all management meetings.
2. Have the environmental manager (if there is one) report directly to top management.
3. Discuss environmental issues and performance while managing by "walking around."

4. Make environmental performance integral to the process for appraising performance.

The first qualitative characteristic for managing the milestones is management commitment. The second is involving the work force.

Involvement of the Work Force

Sustainable performance depends on the full use of the mental resources and commitment of the entire work force. Gaining access to this resource is a function of how much influence people in the company have and how much competence they have. These two variables interact. By extending opportunities for influence, organizations tap more and more of the competencies that members have. As competencies are used, members seek more opportunities to be influential. The more opportunities they acquire to be influential, the more competencies they will develop in order to be more influential, and so on.

Figure 2-1 depicts the relationship between competence and influence. It also suggests how influence and competence are related to commitment and continuous improvement.

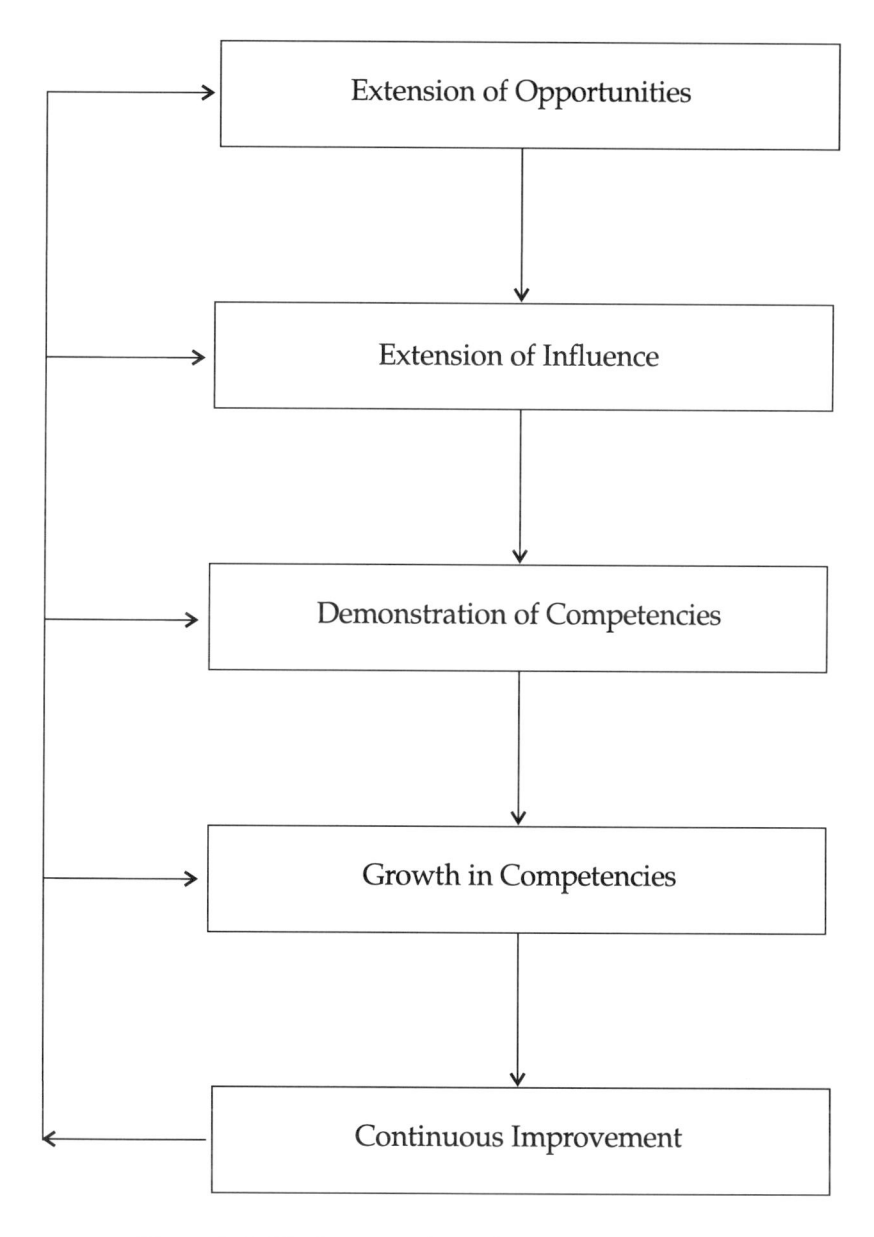

```
┌─────────────────────────────┐
│  Extension of Opportunities │
└─────────────────────────────┘
              │
              ▼
┌─────────────────────────────┐
│   Extension of Influence    │
└─────────────────────────────┘
              │
              ▼
┌─────────────────────────────┐
│ Demonstration of Competencies│
└─────────────────────────────┘
              │
              ▼
┌─────────────────────────────┐
│    Growth in Competencies   │
└─────────────────────────────┘
              │
              ▼
┌─────────────────────────────┐
│    Continuous Improvement   │
└─────────────────────────────┘
```

Figure 2-1. The Process of Involvement and Increased Competent Influence

The easiest and most powerful way to increase the competent influence of the work force is through team formation and development [4]. In high-performing teams, the members challenge one another with alternatives and new ideas, they help one another work toward "stretch" objectives, they tutor and mentor one another, and they help to improve one another's competencies in a variety of ways. Teams provide the best environment for learning and continuous improvement.

Table 2-A suggests areas in which team influence can be extended. The vertical axis lists the kinds of activities through which people can demonstrate influence: by giving information, proposing new ideas, running a test or experiment with a new idea, helping to make decisions, and problem solving. The horizontal axis lists the kinds of team configurations in which influence can be demonstrated: natural work teams, special improvement or problem-solving teams, process-improvement teams, and interface or cross-functional teams.

Table 2-A. Deploying Team Influence

EXTENSION OF OPPORTUNITIES

		Natural Work Teams	Special Improvement Teams	Work-Process Teams	Interface Teams
EXTENSION OF COMPETENCY	Information				
	New Ideas				
	Tests and Experiments				
	Decisions				
	Problem Solving				

Managing the movement of the organization through the milestones requires the full involvement of the entire work force in every step and phase of the process. The organizations that are making the most profound improvements in their environmental performance clearly involve their work forces in the process.

One impressive example of employee involvement comes from the computer industry. Ed Iwasaki of Apple Computer and Paul Russell of Hewlett-Packard met at an environmental packaging conference in Washington, D.C. Afterward, they recruited Ronald Perry from Sun Mycrosystems. The end result was a program called R^3P^2, which stands for "Reduction, Reuse, and Recycling of Protective Packaging." The initial task group now involves the Institute of Packaging Professionals and includes representatives from 100 member organizations. The goal of the group has been to develop common policies and practices among packaging professionals in addressing the environmental impact of the packaging systems they design. A result has been the publication of the *Handbook for Environmentally Responsible Packaging in The Electronics Industry* [3].

Involvement of Other Stakeholders

Sustainable-performance management decisions and actions also involve the organization's stakeholders. The work force is, of course, a stakeholder, but the involvement of employees is so critical at every step that I have treated them separately.

Stakeholders are any individuals or groups who are affected by the organization's environmental performance and who can influence, directly or indirectly, the organization's present or future success.

A list of stakeholders typically includes customers, shareholders, creditors, regulators, suppliers, representatives of the community in which the organization operates, and representatives of environmental and other advocacy groups.

A number of organizations are using advisory groups and panels that represent their stakeholders. These advisory groups serve a number of key functions such as

- Keeping the organization aware of environmental issues that develop.
- Breaking through adversarial relationships between the organization and environmentalists.
- Helping the organization to communicate effectively information about its environmental performance.
- Creating supportive networks that can assist the organization to develop more creative solutions to environmental problems.
- Keeping the organization aware of its image in the local community and further afield.

The kind of talent that can be tapped to serve on stakeholder and company teams is illustrated by a recent move by Dow Chemical. Dow uses a Corporate Environmental Advisory Council to help it to improve its environmental, health, and safety performance. Dow recently added the chairman of the Department of Environmental and Community Medicine from New Jersey's Robert Wood Johnson Medical School to the advisory council.

The full involvement of stakeholders will require the following steps:

1. Conduct a stakeholder inventory and determine exactly who the stakeholders are.

2. Conduct an audit of stakeholder needs relative to the organization's environmental performance.
3. Developing specific strategies and structures for involving the stakeholders in the organization's drive toward SP.

In summary, in order to reach the milestones—or intermediate targets—in an organization's movement toward SP, certain qualitative characteristics of management must be present. These characteristics include demonstrated commitment to SP, involvement of the work force, and involvement of other stakeholders.

The Milestones for Sustainable Performance

The milestones can be reached only with the full involvement of all stakeholders. The SP milestones outlined in this chapter are kept at a fairly general level in order to make them applicable to as wide a range of organizations as possible. At least eight milestones are critical to the successful development of SP. As presented in Chapter 1, these milestones are

1. Sustainable-performance policy statement published.
2. Sustainable-performance baselines established.
3. Initial sustainable-performance training accomplished.
4. Initial improvement projects underway.
5. Development of environmental technologies being supported.
6. Auditing and reporting system functioning.
7. Coalitions exist.
8. Management and human resource systems revised to support sustainable performance.

Milestone One: Sustainable-Performance Policy Statement Published

A policy statement about SP serves the following purposes:

- Clarifies exactly where the organization stands regarding the environment and its business interests.
- Integrates concerns for the environment with the strategic business interests of the organization.
- Focuses the organization's attention on those few interests that are crucial to its success.

Norsk Hydro is Norway's largest industrial company and the fiftieth largest company in Europe. Its Agriculture Group is the largest producer of fertilizers in the world. The U.K. is Hydro's largest single market.

In 1990, Hydro conducted a complete environmental audit of its operations. It hired Lloyd's Register Environmental Assurance to inspect and authenticate the report based on the findings of the audit. Hydro published its Environmental Report in English, Norwegian, and French and used the report to communicate its environmental policy to the world. Its policy is contained in two pieces of the report. First, it is integrated into a statement of the company's "business policy":

> Norsk Hydro is an industrial group, based on the processing of natural resources to meet needs for food, energy, and materials.
>
> Hydro's aim is to initiate development and growth in areas where, by being highly competitive, the company can achieve good profitability.
>
> Hydro intends to satisfy customer requirements by focusing on its markets, and by offering high quality and innovative products.

In all of its activities Hydro will emphasize quality, efficient use of resources and concern for the environment.

Second, the report includes a two-page statement by Hydro's president, Torvild AaKvaag, who has committed Hydro to the following initiatives to improve its environmental performance:

- Developing new, cleaner, and less resource-intensive technological processes.
- Providing information on, and thereby augmenting knowledge of, specific emissions.
- Manufacturing products that are not harmful to the environment.
- Initiating a more environmentally correct use of its products.
- Continuing to work on recycling.
- Actively assisting the authorities in formulating practical, international environmental standards for industry.

A policy statement for SP must meet at least the following criteria. It must

- Contain nothing that is incompatible with the principles and characteristics of SP.
- Be so concrete that there is no doubt about what the organization's values are regarding the environment and how it will conduct its business.
- Clearly tie the business interests of the organization to the environment's best interest.
- Include in specific terms the concept of sustainability.

Northern Telecom already has been mentioned in this book for its positive environmental performance. One reason for its good performance is its well-thought-out environmental policy and the specific objectives that it has developed from this policy. Its policy statement is as follows:

> Recognizing the critical link between a healthy environment and sustained economic growth, we are committed to leading the telecommunications industry in protecting and enhancing the environment. Such stewardship is indispensable to our continued business success. Therefore, wherever we do business, we will take the initiative in developing innovative solutions to those environmental issues that affect our business.

This policy statement meets all the criteria for an SP policy statement. There is nothing in it that is incompatible with the Principles of Sustainable Performance. It leaves no doubt about the organization's valuing of the environment. It ties the organization's business success to environmental performance and expressly includes the concept of sustainability.

The following are additional examples of environmental policy statements that meet one or more of the criteria. The reader is encouraged to judge how these statements might be modified to meet all the criteria for an SP policy statement.

Dow Chemical

> The Dow Chemical Company is committed to continued excellence, leadership and stewardship in protecting the environment. Environmental protection is a primary management responsibility as well as the responsibility of every Dow employee.
>
> In keeping with this policy, our objective as a company is to reduce waste and achieve minimal adverse impact on air, water and land through excellence in environmental control [2].

McDonald's

McDonald's believes it has a special responsibility to protect our environment for future generations. This responsibility is derived from our unique relationship with millions of consumers worldwide whose quality of life will be affected by our stewardship of the environment. We share their beliefs that the right to exist in an environment of clean air, clean earth, and clean water is fundamental and unwavering. We realize that in today's world, a business leader must be an environmental leader as well. Hence, our determination to analyze every aspect of our business in terms of its impact on the environment, and to take actions beyond what is expected if they hold the prospect of leaving future generations an environmentally sound world. We will lead both in word and in deed...[5].

Allied-Signal

It is the policy of Allied-Signal, Inc., to design, manufacture, and distribute all products and to handle and dispose of all materials safely and without creating unacceptable risk to health, safety, or the environment. The corporation will:

- Establish and maintain programs to assure that laws and regulations applicable to its products and operations are known and obeyed;
- Adopt its own standards where laws or regulations may not be adequately protective, and adopt, where necessary, its own standards where laws do not exist;
- Stop manufacturing or distributing any product or carrying out any operation if the health, safety, or environmental risk or costs are unacceptable.

The first milestone that leaders must help their organizations to reach is to develop and publish an SP policy that meets the specific criteria described above. The second milestone is to develop baselines from which improvement objectives can be derived.

Milestone Two: Sustainable-Performance Baselines Established

Baselines are data that represent the organization's performance in a particular area. Benchmarks are baselines that represent what the known best performance in an area is. Baselines provide the organization with information that can be used to solve problems and to develop improvement targets.

The nature of the organization's business will determine the particular kinds of baselines that will be developed and used. The Systems Model for Sustainable Performance provides a basis for determining the general areas where baselines and benchmarks can be created usefully. If we look at the output arrows of the model (Figure 4-1), we can identify these opportunities to gauge environmental performance:

1. Direct outputs to the environment during work processes and as end products of processes. Such outputs include waste emissions, fugitive emissions, pollutants, untreated waste materials (such as building-construction wastes), metal scrap, paper, and plastic.
2. Indirect outputs to the environment through packaging and wastes and pollutants resulting from customer use.
3. Treated and managed wastes through chemicals, biological agents, incineration, deep-well injection, and landfills.
4. Outputs that are reclaimed for reuse.

If we look at the input arrows of the model, we can identify opportunities for improving environmental performance. For example, measuring

1. The waste, pollutants, and toxics created by input from suppliers.
2. The amount of waste created from the output of one step or phase of a process into another step or phase.
3. The amount of water used.
4. The amount of energy used from nonrenewable resources.
5. The amount of energy used from renewable resources.

Specific items that might be monitored on the output side include the following:

- Contaminated water output
- Emission of noxious and toxic gases such as SO_2 and NO_x
- Emission of pollutants such as CO_2
- Fugitive emissions
- Hazardous-waste output
- Hazardous-waste recycling
- Paper recycling
- Plastics recycling
- Product packaging
- Sold-waste output
- Wood recycling

Every organization has an opportunity to conserve energy in order to improve its environmental performance. Energy efficiency comprises a proven set of technologies,

The U.S. increased its economic output by 40 percent between 1973 and 1987 while holding energy demand approximately constant. In a major study undertaken by the United Nations Environment Programme, the following conclusions were reached:

> The steel industry worldwide uses an average of twenty-six gigajoules (GJ) per metric ton of steel produced, whereas the benchmark for the industry is nineteen GJ. The electric arc furnace, for example, uses ten GJ per metric ton. Only 35 percent of steel in the U.S. is made from this process and only 10 percent in the Soviet Union is.
>
> Many opportunities exist in the chemical industry for reducing the amount of energy used; these include upgrading electric-motor efficiency, cogeneration, thermal recompression in evaporation, and automated process control.
>
> In cement making, the dry process uses one-third less energy per metric ton than does the wet process. The wet process, however, accounts for the largest share of production.
>
> The energy use in the paper industry could be reduced with today's technologies by one-third to two-fifths [1].

Minnesota Mining and Manufacturing (3M) Corporation started an energy conservation program in 1973. Through 1989, the program had saved the company $732 million in the costs of fuel and electricity. This savings added $.40 per share before tax on U.S. earnings [8].

Second to energy conservation as a readily available way to improve organizational performance while reducing harm to the environment is managing waste. As with the use of energy, the issue is not that we do not know enough to solve the problem; the issue is that we lack the will and the organization to solve the problem. A general picture of this opportunity is gained by examining the ways in which the U.S., Canada, Germany, and Japan manage waste. Canada recycles about 5 percent of its waste, the U.S. about 10 to 15 percent, and Germany just under 20 percent. Japan, on the other hand, recycles over 50 percent of its waste and uses over 30 percent as fuel to produce new energy.

Sometimes useful baseline data for improving waste management can be obtained from qualified estimates. For example, one computer company began a waste-reduction program by estimating that its packaging materials cost the company $54 million per year, that the average cost per weight of materials was $1 per pound, and that the average density of materials was twelve pounds per cubic foot.

Baselines depend on knowing what the measured environmental performance of an organization is. The most systematic way to develop such information is through audits. Audits and the process of auditing are discussed in Chapter 7.

Milestone Three: Initial Sustainable-Performance Training Accomplished

This milestone, as with all the others, applies to the initiation of the project for building SP into an organization's business practices and work processes. Every milestone, once reached, must be reset in terms of improving it. For example, after the initial milestone for training has been set and as the organization proceeds further toward SP, it will discover the need for new competencies and new training. New milestones will be set.

Training for SP has two major elements. The first element is the generic meaning of sustainable performance, a description of the organization's SP initiative, information about the general SP strategies, and training in the use of SP tools and technologies. This portion of the training also includes (as needed) all the problem-solving and process-improvement tools that currently are associated with TQM. Assuming that every employee can and should be a leader, the general competencies required for the whole work force include

1. The ability to work toward the achievement of SP milestones.
2. A clear understanding of the pressures that are forcing organizations toward sustainable performance.
3. A clear understanding of the meaning of SP, its characteristics and principles.
4. The capacity to help the organization understand and improve the level of response that it makes to the environmental challenge.
5. The capacity to use the general strategies and special tools for SP.

In addition to the general competencies and related training, there is the need for training that is related to the specific kind of business in which the organization is engaged. Specialized training will be required to

- Know relevant laws and how to abide by their enforcement regulations.
- Be able to maintain required documentation.
- Monitor emissions, air quality, and fugitive emissions.
- Conduct audits.
- Monitor ground water.
- Monitor and manage hazardous waste.
- Know how to respond to accidents that affect health, safety, and the environment.

One communications company includes the following special topics in its environmental training program:

- Emergency response procedures
- Handling/disposing of lead batteries
- Underground and above-ground storage tanks
- The halon fire-protection system
- Asbestos
- PCB management
- Disposal of hazardous chemicals

The goal of the training initiative for SP should be to achieve full integration with the organization's ongoing training and development. SP training should start with the first orientation training that new employees receive and, thereafter, have a regular place in all training at all levels.

Milestone Four: Initial Improvement Projects Underway

The fourth milestone in SP is to develop specific improvement projects. Driven by and tied to major improvement objectives of the organization, there are a number of orderly and successful ways to design an improvement project. The following is one example used in TQM projects:

1. Understand the opportunity or problem.
2. Define the specific improvement target.
3. Design strategies to reach the target.
4. Design the data links to track performance and to anticipate necessary adjustments.
5. Design the response process to use data from the data links.
6. Determine how the project will be managed [4].

The key to designing improvement projects is to have data that adequately describe what is going on and then to develop specific improvement targets or objectives. Targets or objectives should be so specific that there is no doubt about the meaning of success. The following are some concrete, environmental improvement objectives from a brewing company:

- Reduce solid waste to landfills by 40 percent by weight by the end of the year.
- Reduce per-unit water use by 10 percent, per-unit wastewater flow by 15 percent, per-unit organic strength by 15 percent, and suspended solids by 10 percent by the end of the year.
- Reduce energy use by 20 percent over the remainder of the decade.

Another set of useful objectives comes from AT&T. AT&T set the following "stretch" objectives, knowing that they may have been beyond reach:

- Phase out CFC emissions from manufacturing operations by 50 percent by December 1991 and by 100 percent by December 1994.
- Eliminate reportable toxic-air emissions by 50 percent by December 1991, by 95 percent by December 1995, and strive for 100 percent by 2000.
- Decrease total manufacturing-process waste disposal by 25 percent by December 2000.
- Recycle 60 percent of paper by December 1994.
- Reduce internal paper use by 15 percent by December 1994 [7].

Key improvement projects should not be limited to those that are clearly coupled to the organization's SP improvement objectives.

Organizations improve their performance through both coupled and uncoupled processes. It is not possible to capture in a set of organizational improvement objectives everything that might be improved or should be improved. One clear lesson from TQM is that team development leads to team-initiated improvement. Of course, nothing should be done that is at odds with organization-wide objectives; but nothing should be done to stifle team creativity and initiative.

Coupled and uncoupled improvement projects can coexist, just as coupled and uncoupled measurement systems can coexist. The common-sense criteria for improvement projects is that

- They always should be focused on improving the quality of performance.
- Priority should be given to projects that support the major improvement objectives of the organization.
- Nothing should be done that is contrary to organizational improvement objectives.

Milestone Five: Development of Environmental Technologies Being Supported

An organization cannot be fully committed to SP unless it also is committed to learning how to do things in radically new and better ways. New technology is an absolute necessity if SP is to develop and flourish.

The development of SP is hindered by the limits of leadership, the limits of economics, and the limits of technology. (Mostly, it is hindered by the limits of leadership—

the commitment and will to change the ways in which the organization does business.)

Large organizations such as Boeing, British Gas, British Petroleum, Chevron, DuPont, Nippon Steel, Norsk Hydro, and Proctor & Gamble support in-house research and development for improved environmental performance. But large or small, every organization can support and use new technology in at least the following ways:

1. Develop in-house environmental technology teams composed of professionals who have access to the publications of and participate in the conferences of their own professional societies. Include the librarian or someone who can access and use environmental and technology data bases.

2. Become active in conferences sponsored by major national and international organizations such as the International Chamber of Commerce, the Global Environmental Management Initiative, the Management Institute for Environment and Business, the Coalition for Environmentally Responsible Economies, and the United Nations Environment Programme. All these organizations have a variety of written resources related to the development of environmental technology.

3. Initiate joint problem-solving activities within the local community and within the organization's own business group.

4. Subscribe to relevant magazines and journals. What an organization will obtain will, of course, depend on the nature of its business. The following are examples of publications that provide up-to-date information about environmental technology and that are relevant to most organizations:

- *Business Horizons*
- *Business and the Environment*
- *Environment*
- *Environmental Business Journal*
- *Environmental Science and Technology*
- *EPA Journal*
- *Garbage*
- *Journal of the Air & Waste Management Association*
- *Nation's Business*
- *Occupational Health and Safety*
- *Pollution Engineering*
- *TAPPI Journal*
- *Technology Review*
- *Water Environment & Technology*
- *Water and Pollution Control*

The fifth milestone of SP is reached when an organization can demonstrate that it is actively involved in supporting and using new technologies to improve its environmental performance. The term "new" will mean one thing in one organization and another thing in a different organization. For example, using computer-assisted heat, ventilation, and air-conditioning control systems to reduce energy use may be new in some organizations. The important thing is that the organization is engaged in using technologies that are new to it.

Milestone Six: Auditing and Reporting System Functioning

The International Chamber of Commerce defines auditing as follows:

...management tool comprising a systematic, documented, periodic and objective evaluation of how well environmental organization, management and equipment are performing with the aim of helping to safeguard the environment... [9].

Chapter 7 presents auditing as one of the important SP tools. This milestone initially has been reached when the following criteria are met:

1. There is an ongoing process of assessing the entire organization's level of SP.
2. Every major organizational unit's environmental performance is regularly audited and documented.
3. A reporting system provides comprehensive information on the organization's environmental performance.

Sustainable performance is data-based performance. Audits serve a large number of purposes. Their main purposes are to

- Determine current levels of performance.
- Determine current and future levels of risk.
- Identify opportunities for improvement.
- Track and document performance over time.

Audits are used for specific purposes such as determining the degree of compliance with environmental law; assessing levels of current risk incurred by present processes and practices; assessing potential risks and liabilities associated with planned purchases of land and facilities; providing proof of due diligence; and evaluating alternative technologies.

Milestone Seven: Coalitions Exist

Coalitions will be discussed again in Chapter 3 as a type of pressure that affects the way in which organizations do business. Coalitions also will be discussed in terms of the Principles of Sustainable Performance.

Forming coalitions is one way in which organizations support the development of environmental technologies. This milestone initially is reached when organizations are active members of coalitions and cooperating groups that include representation from the following:

- Their own type of business (e.g., financial, manufacturing, chemical, transportation)
- National and international business interests
- Local, state, and federal governments
- Environmentalists
- Researchers and experts
- Stakeholders

Participation in joint initiatives to share information and resources, solve common problems, and produce and transfer technology are requirements for SP. SP is communal; every organization exists by the bounty of the environment and within the limits of this bounty. The problems that organizations face cannot be solved with traditional competitive values. The major environmental problems are not limited by proprietary interests. Organizations will solve the major problems together or they will not be solved.

Milestone Eight: Management and Human Resource Systems Revised to Support Sustainable Performance

The final milestone to be reached is that the full range of management and human resource systems in the organization have been revised to support SP. SP is a radically new way of doing business; it represents a major value shift to the Principles of Sustainable Performance. SP becomes more than a project when this final milestone is reached. In fact, reaching this milestone is the point at which the transition takes place from SP being a project to SP becoming a way of life in the organization.

This milestone has been reached when the following can be documented:

- Environmental issues and opportunities are considered, and environmental improvement goals and strategies are incorporated into the organization's strategic plan.
- The organization's accounting systems for resource management include information on water, air, wastes, emissions, and similar factors.
- Accountability for environmental performance has been established for every manager and every member of the work force and is evident in job descriptions and performance-appraisal systems.
- Reward systems are in place to acknowledge special achievements in environmental performance.
- The meaning of quality in the organization has been redefined to include quality of the environment, and total quality management has been revised to include total quality environmental management.

- Hiring and promotion standards and processes include specific mention of skills and performance related to the environment.
- All training, education, and development within the organization is fully congruent with and supportive of the organization's SP initiative.
- All plans and decisions related to every aspect of the organization's performance include proper consideration of environmental issues and opportunities—in marketing, selling, production, engineering, human resource management, and other areas.

Summary

The Sustainable-Performance Management Model (Figure 1-1) communicates three fundamental ideas:

1. There are certain milestones that must be managed in order to initiate and build SP into an organization and these milestones can provide guidance for developing and implementing SP.
2. A number of sources of information can help leaders to manage these elements.
3. It is understanding these elements and using these sources of information that help us to identify the special competencies required for managing SP.

In this chapter, I have described the flow of milestones depicted in the model and the competencies required to manage these milestones. In the next chapter, I will begin to describe the many sources of information that surround the milestones in the model and provide leaders with the competencies they need to manage SP.

References

1. United Nations Environment Programme. (January-February-March, 1990). Energy conservation and Efficiency: A Safe Response to Global Warming and Other Environmental Problems. *Industry and Environment, 13*(1), 6-8.
2. Dow Chemical Company. (n.d.). *Environment, Health & Safety.* Midland, MI: Author.
3. Hewlett-Packard Company. (1992). *Handbook for Environmentally Responsible Packaging in the Electronics Industry.* Palo Alto, CA: Author.
4. Kinlaw, D. (1992). *Continuous Improvement and Measurement for Total Quality: A Team-Based Approach.* San Diego, CA: Pfeiffer & Company.
5. McDonald's Corporation. (n.d.). *McDonald's and the Environment.* Chicago, IL: Author.
6. Norsk Hydro. (1990). *Norsk Hydro Environmental Report.* London: Norsk Hydro Public Affairs.
7. Soderberg, A. (1992). Environmental Policy Deployment: Enriching Our Community. *Proceedings of Corporate Quality/Environmental Management II.* Washington, D.C.: Global Environmental Management Initiative, pp. 91-96.
8. Sombke, L., Robertson, T., & Kaplan, E. (1991). *The Solution to Pollution in the Workplace.* New York: Mastermedia Limited.
9. International Chamber of Commerce. (1991). *ICC Guide to Effective Environmental Auditing.* Paris, France: ICC Publishing S.A.

3

PRESSURES TO RESPOND

Certain major—and possibly nonreversible—changes in the environment caused by humans and their businesses have put our economic and social ways of life at risk. These ways of life include hundreds of conveniences and energy-dependent slaves—all the commercial and domestic machines that we use to produce things, transport things, and increase our comfort. These changes also have put at risk something more fundamental and far more valuable than comfort and convenience. They have put at risk the quality of human life as well as the continuance of life itself.

Our environment is not simply a source of life, it is the source of quality in our lives. Quality largely is experienced as a sense of well-being. Two principle contributors to our sense of well-being are beauty and health.

What happens to beauty in our lives when there are no wildernesses, no lakes, and nothing to see that does not have the mark of humans on it? What happens to this beauty when the sun is obscured by day and the heavens by night by layers of pollutants? What happens when we diminish the richness of life by reducing whole species to perilously low levels and eliminating others forever?

Health, like beauty, is a victim of our own technology. It is a victim of the millions of pounds of pesticides applied each year in the U.S. to crops, lawns, parks, and playing fields. It is a victim of bad air and bad water, of mercury and dioxins, of benzine in gasoline, lead in air pollution, polychlorinated biphenyls (PCBs) in electrical transformers and plastics, and a host of other toxic substances that are routinely present in our everyday lives.

Polluted air, water, and soil are not just threats to health, beauty, and our standard of living, they are threats to life itself. Humans can survive only a minute or so without air, only a few days without water, and only a few weeks without food from the land or sea. We have mounted attacks to our existence by land, sea, and air. We have turned our sources of life into a cosmological dump.

Our Slow Response to the Global Threat

The deterioration of the environment is now worldwide. Changes in one element of the earth's ecology are producing impacts and reverberations—only partially understood—in all other elements. We have moved into a new generation of environmental issues. Local and regional problems of smog, "burning" rivers, contaminated ground water, acid rain, seeping landfills, or even the highly publicized tanker founderings and oil spills are no longer the major threats or issues. The major issues have exploded beyond the limits of single substances such as lead in paints and mercury in water or single events such as toxic spills. The issues now are whether the balance of the fundamental water, earth, and carbon cycles can be maintained; whether the genetic reservoir resident in biodiversity can be protected from further depletion; and whether catastrophic changes in the earth's climate and seas can be averted.

People and businesses have been unnecessarily slow in responding to environmental problems. To some degree, greed, indifference, arrogance, and plain stupidity account for our unwillingness to respond to the environmental challenge. There are, however, two powerful habits of thinking that account for much of our procrastination in responding.

- The first is the habit of perceiving nature as an enemy to be subdued and viewing her as the supplier of endless bounty that must be bent to the will of human beings.
- The second is the habit of viewing the environment piecemeal and its problems in small, separate, easily understood bits.

The Limitless Cornucopia

The most widespread view of nature that has had the most profound impact on our environmental crisis is that nature is a self-renewing resource, an independent provider of limited abundance, a horn of plenty, a "cornucopia." In Western countries, this view has been reinforced by religious belief (especially Judeo-Christian). In the U.S., this view has been further reinforced by the opportunity for colonization and expansion.

According to the Judeo-Christian tradition, God created the first man and the first woman and then said to them:

> Be fruitful, and increase, fill the earth, and subdue it, and have dominion over the fish in the sea, the birds of the air, and over every living thing that moves on the earth. (Genesis 1:28)

Considerable effort is being made these days by theologians and apologists to give a new meaning to this quotation, but it is not clear that such effort will have much effect on a tradition that has placed humans at the top of life's pyramid, "but a little lower than the angels."

Underlying the idea of cornucopia is the idea that nature is an alien force to be conquered. Nature's bounty is not always freely available. Often it must be obtained by force. Early on, the force was physical, but with the rise of science and industry, the force became technological.

The "religious" attitude toward nature received considerable reinforcement in the U.S. from the first Puritan pilgrims and their followers, who moved relentlessly west. The American continent provided an endless opportunity for people to exercise their "god-given right" to cut, kill, dig, plant, burn, and displace. Alexis de Tocqueville, the most famous commentator on American life in the 1800s, described Americans' attitude toward nature thus:

> ...Americans...are insensible to the wonders of inanimate Nature, and they may be said not to perceive the mighty forests which surround them till they fall beneath the hatchet. Their eyes are fixed upon another sight: the...march across these wilds—drying swamps, turning the course of rivers, peopling solitudes, and subduing Nature [8].

To be sure, a logger in the northwestern U.S. who cuts trees to make a living can be expected to have a different point of view on the subject of deforestation from someone in another vocation.

The first reason that individuals and industry have not responded quickly to our environmental crisis is the habit of regarding nature as a boundless and inexhaustible resource that is there for the taking. A second reason for our inaction has been the habit of fragmenting and compartmentalizing the problem.

Fragmentation

The way in which most information about environmental problems is presented in the media leaves the impression that these problems are isolated and not connected to one another and to the whole biosphere. We read about the dangers of some landfill, the toxic residue and emissions at the Love Canal, air pollution in Los Angeles, medical wastes floating ashore at a beach, the ozone hole over Antarctica, and the odyssey of New York's garbage scow without recognizing them as interrelated symptoms of a sick ecological system.

The Problems Are Systemic and Global

The stresses that we are creating in the environment are not the sum of impacts caused by isolated events such as an oil spill, a leaking landfill, or ground-level ozone and nitrous oxide. We are causing stresses to the earth's total ecology.

The closest that concepts such as ecology and ecological systems came to becoming part of popular consciousness was in 1962, with the publication of Rachel Carson's beautiful and disturbing book, *Silent Spring*. Carson explained in her scientific way how DDT could move from one life form in the food chain to another, becoming more concentrated and more lethal as it was passed along. Used as a pesticide on plants, it passed into the soil and percolated into streams, rivers, and lakes. Small organisms passed it to fish, which were consumed by birds. As a result, the reproductive processes of birds were radically altered; their eggs had paper-thin shells, and their offspring failed to hatch or were deformed. Silently and inexorably, DDT passed through the ecological system.

If the natural environment is all that exists that is not made by humans, the ecological systems of this environment are its various living and nonliving elements, their relationships, and their interchanges. An ecological system is, then, the flow of materials or information from the nonliving elements into living organisms and back again.

Ecosystems perform a variety of large-scale, critical functions for which there are no substitutes. These functions, although irreplaceable, have been treated as though they were free and inexhaustible—especially by business and industry. Ecosystems provide the following functions or services from which human life and human productivity are derived:

1. Maintaining the gaseous composition of the atmosphere.
2. Keeping climate changes sufficiently gradual so that life forms can adapt to such changes.
3. Regulating the earth's water cycle.
4. Generating and maintaining soils.
5. Disposing of wastes and cycling of nutrients.
6. Maintaining the global carbon cycle and the nitrogen cycle.

One might think of one's own body as a similar system, made up of organs, skeleton, muscles, and various fluids in which

- All of these exist in an interdependent relationship.
- A change in any one element in the body creates changes in all other parts of the body.
- All of these interact and create something that is a great deal more than the sum of its parts.

Ecosystems, like all systems, have these same characteristics of relationship, interdependence, and of the entity they create being greater than the sum of the parts. Ecosystems also are self-regulating by means of feedback relationships. Take, for example, a greatly simplified ecosystem of hunters, grass, rabbits, and foxes. As hunters reduce the number of foxes, the number of rabbits increases. The more rabbits, the less grass. Decreased grass limits the number of rabbits. Fewer rabbits reduce the supply of food for foxes, and foxes may be reduced to the point of extinction.

The Changes and Issues

Environmental problems are complex and interrelated, and their long-term impact is difficult to predict with certainty. We have reliable information that certain kinds of changes will occur. We do not have precise information about the magnitude of such changes and the timing for them. It is possible to use the uncertainty about the size and schedule for environmental changes as a rationale for doing nothing—or doing only the minimum. However, as time goes on and as more and more information becomes available, even the most skeptical industrialist and the most apathetic individual will recognize that at least the following threats must be averted:

- Global warming and accelerated climate change
- Ozone depletion in the upper ionosphere
- Air pollution and imbalance of life-maintaining gases in the atmosphere
- Contamination of ground-water supplies
- Contamination of the earth
- Pollution and poisoning of the oceans
- Loss of topsoil
- Loss of biodiversity

Commercial organizations may be able to discount the effects of pollution and climate warming. They are not, however, impervious to the multitude of economic and social pressures that will force them to improve their environmental performance or to cease to function.

Organizations that do not accept the absolute severity of what is at stake (or that view the threats to the environment as the imaginings of extremists) will defer action and try to continue business as usual: depleting nonrenewable resources, fouling the water and air, and continuing the ecological destruction that will become self-reinforcing and unalterable. Some of these organizations may wait too long to begin the process of their own renewal. Renewal is not necessary; it is simply the alternative to death.

Growing Pressures to Change

Whether or not organizations respond to what is at stake, they will respond to the pressures that result from the growing recognition of the ultimate stakes. Pressures are the multitude of immediate forces, such as laws, fines, and consumer demands, that will force organizations to move into the environmental age or force them out of business.

The pressures on organizations to respond to the environmental challenge include at least the following:

1. **Compliance.** The increasing number and stringency of laws and regulatory requirements.
2. **Punitive Fines and Costs.** The fines for noncompliance and the costs associated with responses to accidents and disasters are growing in frequency and amount.
3. **Personal Culpability and Imprisonment.** Individuals are being fined and threatened with imprisonment for violating environmental laws, and more such laws are being passed or are already on the books.
4. **Environmental Activist Organizations.** There has been a proliferation of the number of such groups

and their reform agendas at the international, national, state, and local levels.

5. **An Aroused Citizenry**. Citizens are becoming informed through media and other more substantive sources, and they are seeking a variety of channels through which to voice their desires to organzations.

6. **Societies, Coalitions, and Associations**. Professional societies, trade associations, and various *ad hoc* coalitions are making pronouncements and initiating programs that influence organizational behavior toward the environment.

7. **International Codes for Environmental Performance**. The *Valdez Principles* issued by The Coalition for Environmentally Responsible Economies and the *Business Charter for Sustainable Development* created by the International Chamber of Commerce are creating global pressures for environmentally responsible performance.

8. **Environmentally Conscious Investors**. Stockholders are paying closer attention to the environmental performance and rating of corporations. Stockholders include individuals, institutional investors, and managers of retirement portfolios. Organizations' environmental performance and the potential financial risk for poor performance (fines, cleanup costs, litigation expenses) help to determine how attractive their stocks are to investors.

9. **Consumer Preference**. Consumers are looking for green companies and green products and are becoming aware enough to question green hype and green advertising.

10. **Global Markets**. International competition is being carried out within the context of a multitude of environmental laws that will no longer permit companies from developed countries to export their pollution to developing countries.

11. **Global Politics and International Organizations**. A variety of international organizations and forums such as the United Nations' World Commission on Environment and Development, "Earth Summit 92," and the Coalition for Environmentally Responsible Economies exert direct pressure on nations, which invariably affects business and industry.

12. **Competition**. The pressure that stands at the intersection of all others is from the competition and those companies that are moving toward SP, reducing wastes and costs, and finding new, green market niches.

13. **Other Pressures**. At least two other emerging forces will have a strong impact on the way organizations perform in the age of the environment. First, people will prefer to work in organizations with good environmental records. Second, current markets do not reflect the true costs of environmental degradation associated with the business enterprise. In the "full cost pricing" that will come, the prices of products and services will reflect not only production and delivery costs but also the full costs of environmental degradation associated with those products and services.

No one pressure exists independent of the others, and all of them impact on the capacity to compete. It is imperative, however, that organizational leaders understand as much as possible about the nature of each of these pressures in order to help their organizations develop plans and strategies for SP.

Compliance

Federal, state, and local governments are introducing laws that require organizations to manage toxic and solid waste, desist from polluting the air, radically reduce emissions of carbon dioxide, stop emitting chlorofluorocarbons that eliminate ozone from the troposphere, and preserve water resources and their potability. A brief description of the primary U.S. Federal laws is in the Appendix. This list includes the following:

- Clean Air Act
- Clean Water Act
- Resource Conservation and Recovery Act
- Comprehensive Environmental Response, Compensation, and Liability Act ("Superfund")
- Safe Drinking Water Act
- Toxic Substances Control Act
- Federal Insecticide, Fungicide, and Rodenticide Act
- Surface Mining Control and Reclamation Act
- Community-Right-to-Know Act

One phenomenon that appears to be growing is that states in the U.S. are adopting more strict environmental standards than the Federal government requires. Examples are the clean-air standards adopted by California, then by Pennsylvania, New Jersey, New York, and Massachusetts. One implication of this is that automobile manufacturers will now have to build automobiles that meet the higher standards in order to sell them in these states—regardless of Federal standards.

California's Toxics Law applied toxic status to more chemicals in 1989 than the U.S. Environmental Protection Agency (EPA) had done in twelve years under the Toxic Substances Control Act.

New Mexico has passed a law establishing an air-emission-permit system that requires *anyone* who constructs, modifies, or operates a facility whose emissions *may* contribute to air pollution obtain a permit. State and local officials may issue compliance orders and sue in state courts if they see that human health is endangered, *even if the activity does not violate a state rule or permit requirement.* Offenders may be fined as much as $15,000 per day and imprisoned for up to nine years. Those who endanger the health of others may be fined as much as $250,000 [16].

Compliance requirements are affecting not only the ways in which organizations do business, they are affecting decisions about mergers and acquisitions. The challenge to buyer and seller is to keep environmental issues from killing a deal. In one recent acquisition, an energy company discovered that it had also "bought" liabilities for groundwater contamination in the amount of $100 million.

Regulations and compliance with regulations are affecting the capacity to start new businesses and extend the capacities of existing businesses. Lenders to new and old businesses are becoming more and more demanding about the type and extent of assurances on environmental matters before they commit their funds.

There are now over 100,000 environmental regulations in the U.S. at the Federal, state, and local levels. The U.S. Chamber of Commerce estimates that the paperwork and regulatory costs to business will increase by

25 percent in this decade, and the total cost may exceed $600 billion.

The first and most unavoidable reason for organizations to go green is that it is required by law. For the long term, we can expect laws to become more extensive and more rigidly enforced. Organizations that try to circumvent the law will find that violations carry heavy financial risks in fines and cleanup costs.

Punitive Fines and Costs

As regulatory requirements have increased, and as environmental activist groups have pursued polluters more relentlessly with litigation, the costs for noncompliance and for cleaning up after accidents have grown proportionally. The EPA has continued to increase its budget for policing the performance of organizations and levying fines for noncompliance. The costs for not complying with regulations and not taking preventative action to avoid accidental spills and toxic emissions already are greater than the costs of compliance and prevention. The following are some examples.

In February, 1992, the EPA filed a complaint against the chemical systems division of United Technologies, Inc., seeking a fine of $558,000. The EPA charged that the division was burning waste rocket propellant and used solvents in open pits at its facility in San Jose, California [39].

In October, 1992, Triple A Machine Shop, Inc., incurred the largest fine in the state of California's history. The fine, imposed by a San Francisco Superior Court Judge, was levied against the company for dumping contaminated sand, diesel fuel, and dangerous chemicals onto the bay shoreline.

In 1991, the EPA fined two large chemical companies for their failure to comply with the premanufacturing notice (PMN) requirements of the Toxic Substances Control Act (TSCA). The EPA has proposed a $256,000 penalty against Union Carbide Chemical and Plastics Company in Danbury, Connecticut, for manufacturing two chemical substances that were not on the TSCA Chemical Substance Inventory, without first submitting a PMN to the Agency (the proposed fine was cut in half because the company disclosed the violation itself).

Monsanto Company, several other chemical companies, and an insurance company agreed to pay $200 million to the citizens who live near the Brio Refining, Inc., Superfund site, located south of Houston, Texas. This is the largest such settlement reached to date.

The cleanup costs and third-party liabilities associated with catastrophic events such as the Union Carbide spill in Bhopal, India, and Exxon's oil spill in Prince William Sound can carry devastating costs—even for the largest companies. The latest figures for Superfund-related cleanups exceed $500 billion. The ever-mounting costs of the Superfund program have prompted an explosion in the number of lawsuits. A study by The Rand Corporation concluded that as much as 88 percent of total Superfund expenditures went for legal costs associated with either defending environmental-protection policy holders or defending insurers in denial of coverage cases [15].

Rockwell International Corporation pleaded guilty to ten criminal counts of hazardous waste and water violations between 1987 and 1989 at its Rocky Flats nuclear weapons plant. The company agreed to pay a fine of $18.5 million [18].

The owner of a chemical-disposal company in Winter Park, Florida, was ordered to repay the U.S. government

$1.2 million for cleanup of toxic-waste sites. The owner of the company was sentenced to thirteen months in a Federal prison. Companies that paid the disposal company to dispose of their hazardous waste settled with the government and agreed to clean up the contaminated ground at an estimated cost of $5.7 million.

Companies that are in the "environmental" business (e.g., garbage collection, incineration, recycling) can be major violators of environmental regulations and recipients of huge fines. Waste Management, Inc., for example, paid $50 million in fines and related settlements for actions prior to May, 1991, such as unplugging pollution-monitoring devices and disposing of poisonous material by mixing it with oil [41].

The legal liability of the corporation for the environment has been established. This liability is steadily being extended. When a corporation is investigated for a violation, it can expect to be demanded to show that it has an effective environmental policy and plan and that it has educated its work force to carry out the policy and plan.

The vulnerability of corporations for noncompliance and the costs of paying for cleanups are only part of the picture. More and more, the individual employees of organizations can expect to be potential targets for litigation.

Personal Culpability and Imprisonment

There is a movement to increase individual culpability and to make individuals subject to criminal charges, fines, and imprisonment. For example, provisions of the new U.S. Clean Air Act (CAA) make every employee vulnerable for participating in the violation of environmental regulations. Any knowing violation of a CAA regulation or permit standard by an individual is now considered a felony, which is punishable by a fine and up to five years of imprisonment.

The EPA increased its criminal-enforcement activities directed at company officials tenfold between 1982 and 1990. According to a survey in 1991, the U.S. public is in favor of mandatory jail sentences for any decision maker whose organization fails to comply with environmental regulations—if he or she is warned in advance. In the same survey, 45 percent of the respondents strongly favored jailing the CEOs of polluting companies [33].

It is clear that the EPA has been moving away from an earlier role of consulting with companies about compliance to a role of enforcement and criminal prosecution. This movement is highlighted in the growing trend of the EPA and the Occupational Safety and Health Administration (OSHA) to cooperate more closely on field and enforcement activities. If inspectors from OSHA see a violation of EPA regulations, the information will be passed to EPA. Trade organizations are now offering a variety of consultative services to their members to assist them in responding to the increased threat of enforcement and criminal prosecution [7].

The U.S. Clean Air Act (CAA) lists the following violations as criminal conduct:

1. Knowing omissions of material information.
2. Knowing failure to notify or report as required.
3. Knowing alteration or concealment of data or files.
4. Failure to file or maintain documents required by the CAA.
5. Knowing failure to install required monitoring devices.

Bill HR 5305, introduced into the U.S. House of Representatives in June of 1992 and still in committee as of this writing, would bring most environmental statutes up to the standards of the most stringent environmental-crime provisions. The bill would toughen the penalty provisions of several major laws, including the Safe Drinking Water Act; the Comprehensive Environmental Response, Compensation, and Liability Act; the Toxic Substances Control Act; and the Emergency Preparedness Community Right-to-Know Act. The proposed bill would add a "knowing endangerment" provision to all of the listed statutes, authorizing jail sentences and stiff fines for anyone who knowingly places another person in imminent danger of death or serious bodily harm. It also would prohibit the Federal government from awarding contracts to any company convicted of "knowing-endangerment crimes."

States in the U.S. also are passing laws that make individuals criminally liable for acts committed against the environment. Recent legislation passed by the state of Michigan allows citizens to file suit against anyone believed to be seriously contaminating the air, water, or land resources. The Michigan law allows private citizens to take the initiative against major polluters without the delays typically caused when filing through a regulatory agency or bureaucracy.

The first three pressures described stem from the increase in environmental statutes and their accompanying enforcement regulations. These legal pressures have, themselves, often been the result of the actions of individual citizens and environmental activist organizations.

Environmental Activist Organizations

McDonald's fast-food organization was forced by environmental groups to modify its packaging. DuPont was forced to stop manufacturing chlorofluorocarbons (CFCs) by a consumer boycott organized by such groups. The operations of the logging, fishing, and whaling industries have felt the powerful effect of confrontations with such groups.

The number of local, state, national, and international environmental groups now runs into the thousands. A list of just the major ones active in the U.S. is a 200-page book [28]. The growth of the leading environmental organizations operating in the U.S. is displayed in Table 3-A.

Table 3-A. Growth of Membership in Selected U.S. Environmental Organizations

	1970	1990
National Wildlife Federation	2.6m	5.8m
Sierra Club	114,000	645,000
National Audubon Society	105,000	549,000
Wilderness Society	66,000	400,000
Environmental Defense Fund	10,000	220,000
Greenpeace*	6,000	2.4m
Natural Resources Defense Council**	15,000	138,000

*founded 1971 (Greenpeace technically does not have members, only supporters.)
**founded 1970

Environmentalism today is a well-funded, multiheaded juggernaut whose organizations are staffed by bright, hard-working, and often well-paid professionals. DuPont Vice Chairman Elwood Blanchard recently remarked at an environmental-issues conference that the pressure of environmental groups "will continue to be a major force for years to come. We're dealing with bottom-line issues of survival and profitability" [37].

Activist groups now often are included in the category of nongovernmental organizations (NGOs). At a meeting of these organizations in March, 1990, an attempt was made to produce an agenda for the United Nations conference for NGOs planned for May, 1990. A joint resolution called for things such as

- A declaration of environmental rights that would have the same effect as the UN's declaration of human rights.
- A requirement for every country to produce an annual state of the environment report.
- The elimination by all countries of all toxic emissions harmful to human health or the living environment by 1995.
- The reduction of carbon-dioxide emissions by 20 percent from the 1988 level by the year 2000.
- The abandonment of nuclear energy.
- The export and import of wastes to be outlawed immediately.
- All industries to accept "cradle-to-grave" responsibility for all their products [11].

Environmental activist groups are formal examples of an aroused citizenry. Citizens, however, are voicing their displeasure with the environmental performance of organizations in ways other than through formal groups.

An Aroused Citizenry

Local heroes and local causes are increasing. Industrial, energy-producing, and manufacturing organizations are easy targets. NIMBYism (Not in My Back Yard) is requiring companies that hope to expand in settled areas to be squeaky clean. It is demanding that new power stations be avoided, and it is raising the cost and difficulty of disposing of industrial and municipal waste.

The New York Times/CBS News poll regularly asks the public if "protecting the environment is so important that requirements and standards cannot be too high, and continuing environmental improvements must be made regardless of the cost." In September, 1981, 45 percent of respondents agreed with this intentionally extreme statement. In 1989, 79 percent agreed with the statement [26].

A 1991 survey by the American Talk Issues Foundation (ATIF) found that 93 percent of U.S. residents favor the United States' taking the lead in solving global environmental problems. Citizen groups all over the United States promote "Sun Day" to organize grassroots support for a national shift to solar energy [21].

Information and Education

An aroused citizenry typically is an informed one. The Valdez oil spill, the Bhopal toxic-release disaster, and the meltdown at Chernobyl stayed on the front pages of newspapers for days and were prominent in television news programs. Occasional articles about them continue to appear, which seems to indicate that people are concerned about the ongoing effects of such disasters.

Even beyond these spectacular events, the media is including more routine coverage of environmental issues. In the U.S., public television has aired its ten-part program, *Race To Save the Planet*, several times.

The Turner Broadcasting System (TBS) launched a very successful, weekly environmental-magazine series, *Network Earth*, several years ago. Each week *Network Earth* explores a major environmental issue through the eyes of the people who are involved in it. The program includes "local hero" profiles and tells the stories of people and organizations whose public stands on behalf of the environment have inspired others to action. The program recaps the major environmental happenings of the previous week, updates viewers on relevant governmental actions, and provides tips on how to live a more earth-friendly life.

Environmental topics now are included in many grammar school classes and in high school science classes. Colleges and universities are offering more classes on environmental law, environmental science, environmental management, and environmental auditing. The University of Florida, for example, has created a new College of Natural Resources and Environment.

Centers for environmental research and education can be found at a number of universities. The Management Institute for Environment and Business (MEB) in Washington, D.C., is a coalition of academic, government, and corporate resources dedicated in part to fostering environmental education in schools of business.

In 1990, Tufts University formed The Environmental Literacy Institute (TELI) to help faculty members to develop the capacity to include environmental content in the curricula of all major disciplines. TELI also is an excellent example of the kind of cooperation that can be developed among academic resources, business organizations, and government. TELI initially was funded by a two-year grant from Allied-Signal, Inc. Tufts also received grants from the EPA and from Union Carbide. In its first year of existence, TELI developed the capability of twenty-five faculty members to include instruction about the environment into such diverse curricula as economics, mechanical engineering, history, international diplomacy, drama, sociology, and chemistry [35].

Additional examples of environmental-education initiatives involving corporations are as follows:

- Fuji Photo Film U.S.A., Inc., has undertaken a program with the National Audubon Society to provide training and assistance to communities in setting up local recycling programs.
- Toyota Motor Sales, Inc., is cooperating with the National Science Teachers Association in its TAPESTRY Program (Toyota's Appreciation Program for Excellence to Science Teachers Reaching Youth) to promote environmental education by giving grants to secondary-school science teachers.
- The U.S. National Wildlife Federation, in cooperation with sixteen corporations, has a curriculum-development program underway aimed at improving environmental management and policy training in business administration and management schools.
- One of the longest existing educational initiatives is the Alliance for Environmental Education sponsored by Chevron, Edison Electric Institute, The Society of the Plastics Industry, and various other groups. The Alliance produces and disseminates educational materials to schools and the public and operates a network of centers for environmental education throughout the U.S. [17].
- Project Copernicus has been set up by the Conference of European Rectors to introduce environmental training into 400 schools and universities and technical training colleges throughout the European Community.
- The German Environmental Management Association works directly with companies to help workers, supervisors, and managers to become "environmentally intelligent" [22].

The role of the U.S. government in educating citizens about the environment was given a boost with the National Environmental Education Act of 1990. The Act requires that the EPA start an Office of Environmental Education; establish training and grant programs; institute postsecondary environmental internships with Federal agencies; and create a series of national environmental awards programs.

The more informed citizens become, the more they will see that environmental protection and their own self-interests are in complete harmony. Furthermore, they can be counted on to insist that corporations demonstrate their capacity to exist in harmony with the environment. An informed citizenry is an aroused citizenry, and every public and private institution will ultimately bend to its influence.

Whistleblowers

"Whistleblowing"—telling about company practices that are harmful to the environment—may derive from all sorts of reasons other than the desire to right a wrong. However, people who are concerned for the environment can be expected to do some whistleblowing. An employee of Dap, Inc., an Ohio spackle company, told the EPA that Dap had 153 leaky barrels of hazardous waste on site. The employee, after being fired, sued for $3 million for defamation of character and back pay. In 1991, Dap settled out of court with the former employee.

Societies, Coalitions, and Associations

There are two obstacles to general public demand to protect the environment: the first is ignorance, and the second is indifference. However, there is a growing consensus among the people of greatest influence (politicians, business leaders, religious leaders, and professionals) that organizations must start to behave responsibly toward the environment. This consensus is particularly evident in the activities of many types of professional societies, trade associations, and coalitions.

Business and Trade Organizations

Most business and trade organizations are initiating environmental actions within their own communities. The American Textile Manufacturers Institute (ATMI) has developed a ten-point "Environmental Excellence Program." Participating companies will adopt a corporate environmental policy, establish employee and community awareness programs, and conduct voluntary environmental audits [2].

The major plastic producers in the U.S. have formed the Partnership for Plastics Progress. Members include Amoco, Chevron, Dow, DuPont, GE Plastics, Monsanto, Quantum Chemical, and others. The group's charter is to coordinate and improve existing recycling, conservation, and resource-recovery programs; to develop new initiatives; and to do a better job of communicating with the public about its commitment and efforts.

The Institute of Food Technologists (IFT) issued a position statement expressing its support for food packaging that employs integrated waste-management methods, including source reduction, composting, recycling, waste-to-energy conversion, and anaerobic digestion.

Seven U.S. advertising and marketing agencies have adopted the Environmental Initiatives for Advertising Agencies, a set of six environmental principles. The initiatives call for companies to support in-house environmental awareness and education programs, source reduction, pollution prevention, and recycling.

The Chemical Manufacturers Association has developed "Responsible Care" principles that set out guidelines for "product stewardship." The Responsible Care Program is international in scope, largely because the chemical industry is dominated by global corporations. The program already has matured to the point that companies are developing ways to audit their own compliance with the Responsible Care principles [6].

The Computer and Business Equipment Manufacturers Association is cooperating with the EPA to develop guidelines for energy-efficient and environmentally friendly computers. The initiative is called the "Energy Star Computer Program." Participating companies will commit to designing computers that use less energy, supporting programs to reduce energy use in existing products, and displaying an Energy Star Computers logo on products and advertisements.

The Institute of Packaging Professionals (IoPP) has developed environmental guidelines to provide decision makers with a tool that they can use to evaluate and select packaging options. Four of the criteria relate to environmental impact, and the fifth refers to current and future compliance with the law. The criteria are as follows:

1. Source reduction
2. Recycling
3. Degradeability
4. Disposal
5. Legislative considerations

Political and Professional Coalitions

The Coalition of Northeastern Governors (CONEG) is a political coalition that has a reputation for being proactive in developing innovative environmental legislation.

CONEG was formed in 1989. One of CONEG's important achievements was to draft model toxic-materials legislation that subsequently was passed by six states represented in CONEG. CONEG also has established the "preferred packaging" guidelines that follow (in order of priority):

1. No packaging
2. Minimal packaging
3. Consumable, returnable, or refillable/reusable packaging
4. Recyclable packaging and/or recycled material in packaging

These guidelines are included in a variety of proposed legislation in states represented in CONEG.

Various groups of scientists have joined together to issue their own statements and warnings about the environment. The International Council of Scientific Unions (ICSU) has prepared a set of recommendations for governments to influence the level and direction of research required to understand the complexity and precariousness of the "Earth System" [30].

In December, 1992, Baruch Blumberg, James Watson, Hans Bethe, Leon Lederman, Glenn Seaborg, Henry Kendall, and ninety-five other Nobel laureates, along with 1,479 other top scientists from seventy nations signed the document, *Warning to Humanity*. The warning urges the heads of governments to halt air pollution and water pollution, conserve natural resources, stem the loss of wildlife, eliminate poverty, and stabilize population.

The signatories represent more than half of the world's living Nobel Prize winners in science and medicine. The warning states that the earth's ecosystems are under such enormous stress that drastic action must be taken within the next thirty years or it will be too late [36].

Environmentalism is creating new coalitions between politicians and other power blocks. Carl Sagan, the popular astrophysicist, is co-chairman of the Joint Appeal by Science and Religion for the Environment. In 1988 and 1990, parliamentarians, scientists, and religious leaders from 100 nations met at the "Global Forum of Spiritual and Parliamentary Leaders." In May of 1992, the group gathered in Washington, D.C., as the Joint Appeal by Science and Religion for the Environment. Its purpose is to obtain the signatures and support of scientific and religious leaders to declarations such as

> ...we declare here and now that steps must be taken toward: accelerated phaseout of ozone-depleting chemicals; much more efficient use of fossil fuels and the development of a nonfossil fuel economy; preservation of tropical forests and other measures to protect continued biological diversity; and concerted efforts to slow the dramatic and dangerous growth in world population and through empowering both women and men, encouraging economic self-sufficiency....

Industrial Coalitions

The need to find better ways to do business that is friendly toward the environment has created a variety of coalitions among organizations that have been traditional competitors. Such coalitions tend to form around specific environmental issues. As time goes on, these coalitions will exert more influence on the development of new environmental technology. It will become very difficult for organizations that are concerned about their environmental images not to stay aware of current environmental technology or not to participate in these coalitions.

In the United Kingdom, Ford Rover, Vauxhall, Jaguar, BMW, Nissan, and Peugeot Talbot have launched the Automotive Consortium on Recycling and Disposal

(ACORD) to develop environmentally sound strategies for the entire U.K. automotive industry.

In the United States, Chrysler, Ford, and General Motors have formed the Low Emissions Technologies R&D Partnership to coordinate research and development efforts on emissions-control technologies through the exchange of technical information and licensing of breakthroughs. The consortium has the more general mission of making the U.S. automotive industry a "highly coordinated and more powerful resource in the achievement of a cleaner environment while becoming more competitive in world markets" [4].

The top executives of BankAmerica, Pacific Gas & Electric, Chevron, Pacific Bell, Safeway, George Lithograph, and Wallace Computer Services have formed the Recycled Paper Coalition to encourage corporate paper recycling and to stimulate the demand for paper products and recycled materials. These CEOs have committed their companies to implement comprehensive paper-recycling programs in-house and to purchase paper supplies made with a minimum of 10 percent post-customer waste [1].

The pressures on organizations to change to ways of doing business that are more environmentally healthy and sustainable are truly enormous. Trade and professional groups and coalitions represent some sources of pressure. Another type of pressure is that created by various international groups.

International Codes for Environmental Performance

Several international groups have developed "codes," "principles," and "charters" to serve as guidelines for organizations in designing and implementing environmental policies. Among such guidelines are the following:

- The *Valdez Principles* (later renamed the *CERES Principles*) produced by The Coalition for Environmentally Responsible Economies.
- The *Business Charter for Sustainable Development* developed by The International Chamber of Commerce (ICC).
- The *Criteria for Sustainable Development Management* issued by The United Nations Center on Transnational Corporations.
- The *Declaration* of the Business Council for Sustainable Development.
- The *Keidanren Global Environmental Charter* published by Japan.

Copies of the *CERES Principles* and ICC's *Business Charter for Sustainable Development* are found in the Appendix to this book. The Coalition for Environmentally Responsible Economies (CERES) introduced the *Valdez Principles* (later the *CERES Principles*) in 1989; by June, 1992, there were forty-six signatories. The sixteen-point *Business Charter for Sustainable Development*, published by The International Chamber of Commerce, had 242 signatories by the end of June, 1991.

Another group that is exerting pressure on organizations to improve their environmental performance is the Global Environmental Management Initiative (GEMI), which was formed in 1990 by a group of U.S. multinational corporations. Although GEMI has not developed its own environmental guidelines for business, it is one of the primary supporters of the ICC's *Business Charter for Sustainable Development* and has published an *Environmental*

Self-Assessment Program based on the ICC charter. GEMI also has taken the lead in encouraging organizations to make the transition from total quality management to total quality environmental management and has published a *Total Quality Environmental Management Primer*. The primer provides helpful information on applying the principles and strategies of TQM to TQEM [19,20].

As the popularity of international environmental guidelines grows, it will become increasingly difficult for organizations not to endorse one or more sets of guidelines without damaging their images, their attractiveness to investors, and their competitive positions. Evidence of having endorsed such guidelines is already being used as one criterion for judging a corporation's potential for responsible environmental performance—the criteria for evaluating the environmental performance of companies used by the Investor Responsibility Research Center (IRRC) and that of ECO Rating International in Switzerland. Environmentally responsible investors have begun to look at organizations' environmental images before they purchase stock.

Environmentally Conscious Investors

Consumers and environmentalists are also investors. Companies not only must attract people who buy their products, they must also attract investors and please their stockholders.

Investment in green companies and participation in green investment funds is growing rapidly. In the short term, it is unrealistic to expect people to invest in companies only because they are green. But we can expect people to make a choice (each time it is financially sound) for the greener and cleaner companies. We also can expect that major companies will begin to hire third parties to look at their environmental-management systems and will include positive assessments in reports to stockholders.

Michael Deck, a member of the faculty of the University of Toronto's school of management, predicts that environmental records will soon be a standard measure of corporate performance, just as debt levels and earnings already are. A number of groups now provide a specialized service to investors by assessing the environmental performance of organizations. The group that has done the most thorough job in this area is the Investor Responsibility Research Center (IRRC), in Washington, D.C.

IRRC has published a *Corporate Environmental Profiles Directory*, which will be updated annually. Profiles have been developed for all the Standard & Poor's 500 companies. IRRC has developed extensive and rigorous criteria for creating its environmental profiles, including the following standards:

- Environmental achievements and objectives
- Current environmental projects
- Environmental capital expenditures
- Environmental staff
- Environmental policy
- Environmental auditing scope
- Environmental insurance availability
- Number of Superfund cleanup sites

IRRC's evaluation includes the use of two indices: *Emissions Efficiency Index* and *Compliance Index*. These provide standardized indicators of corporate environmental performance [24]

Investment in socially responsible stocks and mutual funds grew from $40 billion in 1984 to $625 billion in 1991. Moreover, socially responsible investing (SRI) can be as profitable as any other sort of investing. In 1990, the top-performing balanced mutual fund, Pax World, was an SRI fund. At the end of 1991, the average five-year return of the six publicly available, socially responsible equity funds was 10 percent higher than the average return of all equity mutual funds during that period [17].

The interest in green investing has been growing for some time and is proving to be an attractive alternative to investors. For example, New Alternatives is an environmental fund started in 1982. The company invests in companies such as Energen and Burlington Resources, two firms that produce methane gas for use by utilities, and in Chesapeake, a paper-recycling firm. The assets of the fund in 1982 were $100,000. Today the fund's investments are over $12 million, and, in 1989, a share in the fund increased by 26 percent [31].

Marc Epstein, of the Syms School of Business, conducted a survey of shareholders and determined that shareholders want corporations to direct more money toward cleaning up plants, stopping pollution, and making safer products. Investors want companies to spend money on these areas rather than declaring higher dividends [32].

All companies whose stocks are publicly traded are influenced by the perceptions that investors have of these companies. One investment group published a report that indicated that the environmental liabilities of companies will make them credit risks [32]. Large financial obligations incurred because of the past generation of wastes, environmental disasters, and the financial costs involved in retrofitting to meet regulatory requirements will come under increased scrutiny to determine credit eligibility.

Investors are a special form of consumer. The general consumer is the one who purchases products in stores.

Consumer Preference

The buying habits of people are changing. Joel Makower writes in *The Green Consumer* that every purchase is a vote for or against the environment [14]. The J. Walter Thomson advertising firm revealed a 400 percent rise in the introduction of products that featured green advertising between 1989 and 1990 [9].

The power of an aroused citizenry is most obvious in its purchasing power. A study of the Food Marketing Institute found that 50 percent of the shoppers surveyed said they would switch to a supermarket that promoted environmentally friendly products.

An indication of the degree to which consumer habits are changing is the effort now being made by producers and retailers to show that they are "green." When John Elkington and Julia Hailes published the *Green Consumer's Supermarket Shopping Guide* in 1988, it reached the top of Britain's best-seller lists by its fourth week [13].

One of the oldest "green" products is recycled paper. Printers report that the demand for recycled paper is up over 15 percent from a few years ago — even though recycled paper stills cost from $5 to $10 more per 1,000 sheets than virgin stock. One entrepreneuring printer uses a clever promotional pitch: Inkwell of America, in Ohio, offers customers who order jobs on recycled paper the bonus of having a tree planted in their name [25].

Ecolabeling

The number of people worldwide who indicate a preference for environmentally friendly products has increased drastically over the past few years. Ecolabeling of one kind or another exists in at least half a dozen countries, and others are in the planning stages.

As early as 1977, West Germany introduced its Blue Angel ecolabeling program. By 1989, more than 3,000 products made by 600 companies were authorized to carry the label.

Japan initiated its "Eco-Mark" product labeling program in 1989. The Eco-Mark can be used not only to label products but also to promote environmental actions such as bottle and can recycling.

Canada has an "Ecologo" as part of its Environmental Choice Program. To date, qualifying criteria have been developed for thirty-four product categories, and 650 products from 120 companies are authorized to display the environmental logo.

The Nordic Council of Ministers has planned a "White Swan" ecolabeling program. This program will be binding on products produced or marketed in Norway, Sweden, Finland, and Iceland.

In the U.S., one nonprofit organization has initiated *Green Seal* and a similar organization has proposed *Green Cross*. For a product to receive full environmental certification from the Green Cross Certification Company, it must undergo a total-product assessment of the product, the packaging, and the manufacturing process. Green Seal, Inc., awards its seal to products that are judged to reduce waste, from manufacturing to disposal.

Lister Butler, a New York City packaging-design firm, has developed an environmental-design evaluation process called EDEX that analyzes the efficiency and effectiveness of the packaging material and whether emerging legislation is expected to affect that material [29].

Surveys

Gallup surveys indicate that 75 percent of U.S. consumers include environmental concerns in their shopping decisions [27]. Two-thirds of U.S. consumers would consider switching to a different brand of small appliance if their favorite choice was not packaged in recycled or recyclable materials [2].

According to a 1991 survey conducted by Golin/Harris, adults in the U.S. are willing to pay 25 percent more to decrease air pollution, 50 percent more for garbage disposal, and $1,000 more for a smog-free car. A majority would support mandatory household recycling, and 76 percent would offer tax incentives to businesses to encourage environmentally responsible behavior [1].

Most pollsters agree that green consumerism is on the rise, but people may be more willing to *say* they are green than they are to *act* in their shopping behaviors. A *Wall Street Journal*/NBC survey in 1991 found that eight out of ten U.S. residents regard themselves as "environmentalists"; half of those said they were "strong" ones. What emerged from the survey, however, was that environmental concerns are actually taken into consideration along with more traditional ones such as price and convenience. The president of a New York market-research firm put it this way:

> ...there is a minority who are passionately concerned and will inconvenience themselves to do good environmentally—paying more and changing habits or giving more thought to daily practices [38].

In the extensive research and survey work of John Elkington and his colleagues at SustainAbility, Ltd., there is clear evidence of the rise and growing strength of the green consumer. At the international level, the popularity of their books—such as *The Green Business Guide, The Green Consumer's Supermarket Shopping Guide,* and *The Green Consumer*—is testimony to the growing power of the green consumer [12,13,14].

In the 1990 and 1991 Greenworld Surveys conducted by SustainAbility, Ltd., the trend toward green consumerism is documented as an international phenomenon. It is growing at uneven rates, but it is clearly growing [10,11].

Even though there is some question about the exact correspondence between what people say and what they do, there is no doubt about the general trend toward green consumerism. There is concurrence among marketing groups that it will be difficult to sell any product in the year 2000 that is not environmentally friendly.

Green Does Not Always Mean Informed

Green-minded consumers are not, of course, always informed ones. When popular opinion forced McDonald's fast-food restaurants to switch from plastic "clamshell" hamburger boxes to paper wrappers, it was not because the switch represented a sound environmental decision. McDonald's had researched the problem and concluded several years ago that styrene packaging actually was more recyclable than paper.

There is a widespread misunderstanding among consumers about the relative merits of paper and plastic. The common wisdom is that paper is degradable. The regular practice of supermarkets to offer the customer "paper" or "plastic" reflects the belief that paper will biodegrade in landfills but that plastic will not. Few consumers who choose paper at the checkout counter know that nothing degrades easily in an airtight landfill, whether it be paper, plastic, or something else. By certain criteria, in fact, plastic is more environmentally friendly than paper.

A 1990 study by the Council for Solid Waste Solutions (CSWS) found that polyethylene bags produce 74 to 80 percent less solid waste, 63 to 73 percent fewer emissions, and 90 percent fewer waterborne wastes than paper bags.

Green Marketing and Advertising

Green marketing is business' response to the green consumer. Environmental labels will, on occasion, be used to defraud the public, and there will be no shortage of misleading claims as businesses compete for the environmentally concerned consumer.

Some companies have begun to use nonspecific and occasionally misleading terms such as "environmentally friendly," "recycled," degradable," "biodegradable," "photodegradable," and "compostable" to describe their products. However, many companies and consumers are becoming aware of the guidelines that are proposed in *Green Report II*.

Green Report II was the work of a coalition of state attorneys general from California, Massachusetts, Michigan, Minnesota, New York, Texas, Utah, Washington, Wisconsin, Florida, and Tennessee. The task force proposed guidelines for green advertising to be followed until the U.S. Congress or one of the regulatory agencies produces mandatory green-advertising regulations.

The task force of attorneys general has had considerable impact beyond the issuance of its green report. It compelled American Enviro Products, Inc., the manufacturer of "disposable" diapers to withdraw its deceptive claim that its product was biodegradable in landfills and forced the firm to pay a $5,000 settlement to each of ten states. In July, 1991, the attorneys general cited the Alberto-Culver Company for its claims that the hair sprays it produced were "ozone-friendly." The company paid $5,000 to each of ten states and withdrew its advertising.

The U.S. Federal Trade Commission has become involved with the issue of green advertising. It currently has approximately thirty investigations underway, and it has reached settlements with a number of companies for making misleading environmental claims. Legislation called the "Environmental Marketing and Claims Act" has been introduced. The act would require the EPA to establish uniform and accurate standards for environmental marketing claims.

One of the most reported incidents of questionable green advertising concerned the Hefty trash bag produced by Mobil Corporation. The company's claim that the bag would degrade in sunlight was disputed by Greenpeace. Six states sued Mobil for making false claims, and Mobil agreed to withdraw the claim and to pay $150,000 to each of the states.

Green marketing is not a fad. U.S. respondents now rank environmentalism more important than nuclear safety, national defense, food, or safety [5]. Green marketing will play a critical role in communicating to the public what it needs to know about products and their relation to the environment. In the future, it may do more than any government or environmentalist group to enlighten the public about the environment and its protection.

The interactive feedback processes between consumer practices and what business does are extraordinarily powerful. Most Western nations have consumer cultures; when we change what we buy and how we buy it, we will change what we produce and how we produce it.

Thus far I have discussed the pressures coming from the need for compliance; punitive fines and costs; personal culpability and imprisonment; environmental activist organizations; an aroused citizenry; societies, coalitions, and associations; international codes for environmental performance; environmentally conscious investors; and consumer preference. In the next two sections, I will discuss the pressures that are coming from global markets and global politics.

Global Markets

It is becoming less possible for businesses and industries to export pollution. In the future, countries such as Romania and Mexico will not be able to use their more lax environmental regulations to attract investment from developed countries.

General Motors has decided to relocate its truck-assembly plant from Mexico City because of the new laws on air quality. Companies such as GM have had to face the hard choice of retrofitting their antiquated factories with expensive emission-control equipment or moving their plants. In the same city, the government shut down one U.S. company for two weeks because of severe problems with air pollution. Cummins Engine Company left Mexico City early in 1992 because of the prospective costs to reduce the emissions of its plant [40].

Exporting companies are now finding that the countries that buy their exports may have stricter environmental laws and enforcement than does the country of the exporting company.

Countries are acting both independently and in coalitions to enact a wide range of laws and regulations that require compliance from organizations that do business globally. It is difficult and confusing to manage the often different requirements represented by these laws. The requirements of the most strict country in which a segment of a multinational business is located ultimately will determine how business is done in all the other countries. For example, the Germans are taxing plastic bottles, and the Swiss have outlawed metal containers. To do business in these countries, organizations must comply with their standards.

Another example of the kind of environmental regulations that competitors in a global market face is Germany's new packaging law. As of December 1, 1991, Germany requires that producers and retailers take back or recycle packaging used to ship goods or arrange for someone else to do it. The goal is to collect 80 percent of packaging wastes by 1995. If these goals are not met, the government plans to require up to $.30 deposit per item on all types of packaging.

Global competition in the environmental arena also is becoming more of a factor for international organizations. It is to the advantage of truly green companies to drive the dirty companies from the market and to take their market share. One way to do this is to join forces with the environmentalists and ensure that all environmental regulations are strictly enforced.

The European Community (EC), by its environmental rules, will have a powerful impact on the ability of any non-European company to do business in Europe. The EC has an avowed purpose of enacting environmental rules that reflect the highest standards of any member of the community. The EC Commission (EC's executive body) has put forth an ecolabeling proposal that will have enormous impact on which products can enter any EC nation. The proposed ecolabel would be granted only to products that satisfy very stringent criteria such as life-cycle analysis; use of natural resources; energy consumed; possible damage to air, water, or soil; solid and toxic waste produced; and noise.

The changing rules in the global market are having a dramatic effect on organizations that want to compete in or to maintain plants in other countries. Another pressure of global dimension is that which is exerted through a variety of international agreements.

Global Politics and International Organizations

The environment no longer is a local or national issue. National boundaries are no barriers to acid rain and the transport of other toxics by wind currents in the atmosphere. Romania's chlorine emissions drift across the Danube into Bulgaria. Germany's sulphur emissions from burning lignite in its power plants kill forests in Czechoslovakia. The U.S. exports acid rain to Canada.

International meetings on the environment have resulted in a number of conventions and agreements that determine how waste and pollutants will be managed. One example is the Montreal Protocol to eliminate chlorofluorocarbons. Others are the Convention on Civil Liability for Damages Caused During Carriage of Dangerous Goods by Road, Rail and Inland Navigation Vessels and the London Convention on the Prevention of Marine Pollution by the Dumping of Wastes and Other Matter. The United Nations Conference on Environment and Development (the self-styled "Earth Summit") is a more recent example of the global meetings on the environment that have taken place in the last several years.

The greening of business is an international phenomena. Among the conclusions reached by SustainAbility, Ltd., in its 1990 *Greenworld Survey* of twenty-seven countries were these:

- The environmental movement is growing rapidly throughout the world.
- The global public will give increasing support to environmental protection.
- Environmentalism will be the most dominant political and economic movement globally by the year 2000 [11].

Various international organizations are now attempting to stimulate and support a global response to the global environmental challenge. Many of these organizations benefit from the significant involvement of a number of international corporations such as British Gas, DuPont, Dow, Proctor & Gamble, and British Petroleum. All the organizations are resources for information on sustainable performance and are described in the Appendix. Some of the more important ones are listed below:

- The Business Council for Sustainable Development
- The Coalition for Environmentally Responsible Economies
- The International Chamber of Commerce
- The Global Tomorrow Coalition
- The Global Environmental Management Initiative
- The Responsible Care Program of the chemical industry
- The United Nations Environment Programme
- The World Environment Center
- The World Resources Institute
- The World Wide Fund for Nature

The pressures working on organizations and forcing them to respond to the environmental challenge are many and varied. All of them, however, intersect at one point: competition. The way that organizations respond to them will determine their capacity to compete.

Competition

The most powerful pressure working on organizations and the one that will force them to embrace sustainable performance as a way of life is competition in the marketplace.

Only those organizations that learn to respond creatively to the pressures listed in this chapter can expect to survive. *Green* business is just *good* business. Soon it will be the only way of doing business.

The capacity to compete in the marketplace is the pressure that surrounds all the other pressures. The capacity to compete will not come easily. The capacity to compete will be determined only by organizations' making their current services, products, and processes cleaner. It also will be determined by how quickly organizations are able to create new processes, create new products, and develop new opportunities in the emerging environmental market.

Competitive and Green

Concern for profitability and competitive position are already leading organizations to improve their environmental performance (quite apart from any concerns they may have for life and the quality of life). These improvements have helped companies to save money, to produce better designs, to discover better means of production, to improve their products, and to produce new goods and claim new niches in the marketplace [23].

In a review of five-hundred case studies reported in a joint study by three major international, environmental agencies, it was found that organizations that reduce waste and prevent pollution achieve the following "bottom line" benefits [44]:

- Lower costs of raw materials
- Lower energy costs
- Lower waste-disposal costs and reduced dependence on waste-treatment and disposal facilities

- Reduction or elimination of future liability for cleanup of, or contamination by, buried wastes
- Fewer and lesser regulatory complications
- Lower operational and maintenance costs
- Lower employee, public, and environmental risks and expenses, both present and future
- Reduced liability insurance and costs
- Better employee morale, productivity, and product quality

Competitive success obviously is related to image, and image is a function of several pressures that have been discussed in this chapter. Another way that the image of a company can be enhanced is by the receipt of a prestigious award.

Environmental Awards

Existing organizational awards for environmental performance that have been identified by the Global Environmental Management Initiative are

- Honor Role Awards given by the National Environmental Association to those companies that have found a balance between environmental needs and economic reality. Previous winners of these awards have included Anheuser-Busch Company for water conservation in the Los Angeles area; AT&T for the elimination of CFC emissions; Heinz for its ENVI-ROPET plastic ketchup bottle; and Valvoline for its used oil-bank recycling program.
- The National Recycling Coalition gives recycling awards in ten categories. EcoSource of Green Bay, Wisconsin, was a winner for recycling innovation,

and McDonald's won an award for its corporate leadership.
- The National Wildlife Federation makes an annual presentation of the Corporate Conservation Council Environmental Achievement Award to a U.S. business for exceptional accomplishments in undertaking a conservation project. The Wisconsin Electric Power Company won the award for its Appliance Turn-In program to remove inefficient appliances from use and to recycle the raw materials. 3M won for its Pollution Prevention Pays program. The IMC Corporation won for converting 525 acres of strip-mined uplands into a diverse wetland system.
- The U.S. Environmental Agency sponsors the Administrator's Award. Awards are made in nine categories. One of the awards is made to a small business, and another is made to a large business.
- The White House Council on Environmental Quality gives the President's Environment and Conservation Challenge Awards in four different categories to individuals, groups, or organizations. These categories are Partnership, Environmental Quality Management, Innovation, and Education and Communication.
- The World Environment Center sponsors an annual presentation of The Gold Medal Award. The award honors companies for international, corporate, environmental achievement. Candidacy for the award is limited to corporations with industrial and/or processing functions that are a substantial part of their operations outside the headquarters country. Past recipients include Rohm and Hass, IBM, Dow Chemical, and British Petroleum.

It is unlikely that I have identified all the pressures that are pushing organizations toward improved environmental performance. The list, however, is sufficient to show just how impossible it is becoming for organizations to conduct business without making the environment a full-fledged senior partner. There are, however, some additional pressures that have just begun to put in an appearance, but which certainly will have more impact in the future.

Other Pressures

At least two additional pressures are emerging that will contribute to forces that will compel organizations to transform their business practices to respond to the environmental challenge. These additional pressures are

1. Qualified people will exercise their preference to work in organizations with good environmental records.
2. Organizations will be forced to cope with "full-cost pricing."

It is not clear at this point just how important these pressures will be, but organizations should, at least, begin to anticipate their potential power.

Employee Preference

In several surveys, people have clearly indicated that—given the choice—they would elect to work for an organization or company that has a strong environmental record. Organizations that want the best people should expect to have the best environmental records. A survey by McKinsey & Company in 1991 of 400 senior executives of major companies worldwide found that these executives believed that "organizations with a poor environmental record will find it increasingly difficult to recruit and retain high-caliber staff" [42].

An enlightened and aroused citizenry also means an enlightened and aroused work force. The environment is not an issue that can be left at home. The topic is becoming too public, too popular, and too pressing to be forgotten for very long by anyone. The role of the good citizen is being rewritten. People are not as likely to select a work place in which smoking is permitted; they are less likely to select a work place in which recycling is not practiced. Moreover, if people have the alternative, they will not work for known polluters. We know from several surveys that, given an alternative, people will elect to work for organizations that demonstrate strong environmental values and that have successful environmental-improvement programs.

It is likely that many employees are ahead of their organizations' executives in their commitment to the environment. Edgar Woolard, Chairman of DuPont, in discussing his own company's movement toward environmental responsibility, remarked

> I have observed within DuPont that the people who work for our companies are already committed. Employees tend to share the same environmental expectations as many in the general public. They do not need motivation; they need empowerment by management to do what needs to be done [43].

During the last twenty years or so, the author has conducted a large number of management-education programs that involved senior managers and executives. Until the past two years, these leaders rarely had to respond to questions about their organizations' environmental performance and plans. Now they are always presented with a wide range of questions about the environment. Organizational performance and environmental performance are now inseparable issues.

Full-Cost Pricing

Economists are proposing that "full-cost pricing" be added to the balance sheets of companies. The theory is that organizations will make prodigal use of environmental resources for which they are not charged or are able to use at below-market prices because of subsidies. Surveys made by the Organizations for Economic Cooperation in 1987 and 1991 showed that the international interest in the value of using various forms of full-cost pricing was rapidly increasing in Europe, Canada, and the U.S. [34].

At present, many governments distort the real cost for producing various products and services because of incentives and subsidies. In many countries, for instance, energy is subsidized through supports for coal production and electricity-tax exemptions. In others, such as the U.S., farmers are subsidized to overproduce and bring to market an excess of foodstuffs.

One abusive and environmentally deadly practice of subsidy in the U.S. has to do with cows grazing on public lands. Cows range over 70 percent of the Federal land in the western part of the U.S., under a leasing program that does not pay for itself and which requires tax subsidies each year to produce about 3 percent of the country's beef. The constant grazing makes the rangeland barren. The cattle trample on the nests of wildfowl, pollute the streams with their feces, and turn natural wetlands into gullies of baked earth.

We can anticipate that as the global economy becomes the global environmental economy, countries will insist that they compete on a "level playing field" and that products and services reflect their total costs—including their cost to the environment. Industries that have benefited from preferential government support can not only expect their subsidies to be removed, they can expect to be taxed for the amount of pollution they produce and the cost of future problems they may have created through current actions.

Summary

In this chapter, I have outlined some of the more powerful pressures that are now bearing on organizations throughout the world and on U.S. organizations in particular. All these pressures have economic implications and all of them threaten the success and even the survival of organizations. All these pressures ultimately will converge in the marketplace, and only those organizations that use them to improve the quality of their performance will profit and survive.

Sustainable performance describes the way that organizations will be doing business in the environmental age. The first source of information for managing the milestones of sustainable performance is the set of pressures that are forcing organizations to respond to the environmental challenge. A second source of input is the understanding of SP itself. In the next chapter, I will develop the full meaning of SP.

References

1. *Business and the Environment.* (July 6, 1991). p. 6.
2. *Business and the Environment.* (November 8, 1991). p. 4.
3. *Business and the Environment.* (December 6, 1991). p. 2.
4. *Business and the Environment.* (June, 1992). p. 7.
5. *Business Ethics.* (July/August 1992). p. 30.

6. Chynoweth, E., & Heller, K. (June 17, 1992). BP Baglan Bay (Wales Petrochemicals Complex and Responsible Care Program). *Chemical Week*, pp. 28-29.
7. Consoli, J. (June 22, 1991). Get Your Act Together. *Editor & Publisher, 124*, pp. 7-8.
8. de Tocqueville, A. (1835, 1840). *Democracy in America* (1952 Reprint). London: Oxford University Press.
9. *Earth Island Journal.* (Fall 1991). p. 25.
10. Elkington, J. (1991). *The Corporate Environmentalists: A Report on the 1991 Greenworld Survey.* London: SustainAbility, Ltd.
11. Elkington, J. (1990). *The Green Wave: A Report on the 1990 Greenworld Survey.* London: SustainAbility, Ltd.
12. Elkington, J., & Knight, P., with Hailes, J. (1991). *The Green Business Guide.* London: Victor Gollancz, Ltd.
13. Elkington, J., & Hailes, J. (1988). *The Green Consumer's Supermarket Shopping Guide.* London: Victor Gollancz, Ltd.
14. Elkington, J., Hailes, J., & Makower, J. (1990). *The Green Consumer.* New York: Penguin Books.
15. *Environmental Business Journal.* (July, 1992). p. 8.
16. ES&T Currents. (June, 1992). *Environmental Science & Technology, 25*(6), 71.
17. *Financial Planning.* (September, 1992). pp. 25-27.
18. Rockwell to Pay Rocky Flats (March 27, 1992). *Florida Today*, p. 18C.

19. Global Environmental Management Initiative. (1992). *GEMI at Two, 1991/1992 Annual Report.* Washington, D.C.: Author.
20. Global Environmental Management Initiative. (1992). *Total Quality Management: The Primer.* Washington, D.C.: Author.
21. Henderson, H. (1991). *Paradigms in Progress.* Indianapolis, IN: Knowledge Systems.
22. Industry Initiatives in Achieving Ecologically Sustainable Industrial Development. (1991). *Proceedings of the Conference on Ecologically Sustainable Development.* Vienna, Austria: United Nations Industrial Development Organization.
23. International Institute for Sustainable Development. (1992). *Business Strategy for Sustainable Development.* Winnipeg, Manitoba, Canada: Author.
24. Investor Responsibility Research Center. (1992). *Tracking Business and the Environment.* Washington, D.C.: Author.
25. *Jacksonville Business Journal.* (November 15, 1991). p. 8.
26. Kirkpatrick, D. (February 12, 1990). Environmentalism. *Fortune*, pp. 44-52.
27. Kleiner, A. (July/August, 1991). What Does It Mean to Be Green? *Harvard Business Review*, pp. 38-47.
28. Levine, M. (1991). *The Environmental Address Book.* New York: Putnam Publishing Group.
29. *Management Review.* (June, 1990). pp. 18-23.
30. Marton-Lefevre, J. (June, 1992). Mobilizing International Science. *Environmental Science & Technology*, 25(6), 1085-1087.

31. Newman, S. (June, 1990). Return to Vendor: Aluminum Recycling. *Management Review*, p. 1081.
32. Minnow, N., & Michael, D. (Summer, 1991). The Shareholder's Green Focus. *Directors & Boards*, pp. 35-39.
33. Roper Organization, Inc. (June, 1991). *Environmental Protection in the 1990's: What the Public Wants.* Presentation to the U.S. Environmental Protection Agency.
34. Schmidheiny, S. (1992). *Changing Course.* Cambridge, MA: MIT Press.
35. Smart, B. (1992). *Beyond Compliance.* Washington, D.C.: World Resources Institute.
36. Take Care of Earth, Scientists Tell Leaders. (November 28, 1992). *The Virginian Pilot*, p. A-3.
37. *USA Today.* (May 15, 1991). p. B3.
38. *The Wall Street Journal.* (August 12, 1991). p. B2.
39. *The Wall Street Journal.* (February 5, 1992). p. B4.
40. *The Wall Street Journal.* (March 30, 1992). p. A7.
41. Waste Disposal Giant, Often Under Attack, Seems to Gain From It. (May 1, 1991). *The Wall Street Journal.* p. 1.
42. Winsemius, P., & Guntram, U. (March-April, 1992). Responding to the Environmental Challenge. *Business Horizons.* pp. 12-16.
43. Woolard, E. (November 19, 1990). A Sustainable World. *Chemistry & Industry*, pp. 738-739.
44. World Conservation Union, The United Nations Environment Programme, World Wide Fund for Nature. (1991). *Caring for the Earth.* Gland, Switzerland: Author.

4

CHARACTERISTICS OF AND A SYSTEMS MODEL FOR SUSTAINABLE PERFORMANCE

The new way of doing business toward which all organizations are moving is sustainable performance. In Chapter 3, I suggested that this movement is now inevitable because of the pressures that are creating the requirement to change from a philosophy of total quality management to one of total quality environmental management.

In order to manage the progress of an organization through the milestones for sustainable performance (Figure 1-1), leaders must understand the full meaning of SP. In Chapter 1, I provided some examples of SP, along with the following short definition:

> Sustainable performance is the evolution of organizations into wealth-producing systems that are fully compatible with the natural ecosystems that generate and preserve life.

The purpose of this chapter is to develop more fully the meaning of sustainable performance and to show how an understanding of SP is critical in managing the milestones that lead to it. Specifically, the objectives are to

1. Describe the distinguishing characteristics of SP.
2. Present an open-systems model of organizational performance (the Systems Model for Sustainable

Performance) that properly takes the environment into account.

Sustainable Development and Sustainable Performance

The concept of "sustainable development" entered the popular vocabulary through the work and publications of the United Nations World Commission on Environment and Development (WCED). In 1987, the WCED published its report, called *Our Common Future* [15]. (The report is most commonly referred to as "The Brundtland Report," after the Commission's Chairman, Gro Harlem Brundtland, Prime Minister of Norway.) The report firmly established the idea of sustainable development as the concept that would dominate economic and business policy for future years.

The WCED has defined sustainable development as

> ...development...that meets the needs of the present without compromising the ability of future generations to meet their own needs.

There have been a number of initiatives to operationalize sustainable development within the business community [4,9,12]. The following elements are common to most definitions of sustainable development:

1. **Equity.** This means current equity among all the earth's peoples, in that they all have access to the chance to improve their economic well-being. It means intergenerational equity, in that all future generations have access to an improvement in their economic well-being that is equal to that of current generations.

2. **Stewardship.** This means that the manufacturing, financial, agricultural, and building processes of

development are undertaken in ways that demonstrate full stewardship of what is used and what is produced. It means that all processes, products, and construction are designed to result in the least possible damage to the environment and the earth's underlying ecosystems.

3. **Limits.** This means that development must be undertaken within the known or probable limits of the earth's nonrenewable resources and the limits that its ecosystems can tolerate from human intervention. It means that the use of nonrenewable resources must be planned for the projected period of human need and that any further elimination of living species by human action must be halted.

4. **Communal.** This means that the damage to the earth's environment and ecosystems is not bounded by national or geographic boundaries; only quite broad or global perspectives and broad-based cooperation can repair the harm that already has been done and ensure environmentally safe development in the future.

5. **Systemic.** This means that development must proceed with full awareness of the interrelationships among all natural ecosystems and all human activity.

Organizations that are leading the movement for business involvement in sustainable development are the International Institute for Sustainable Development (IISD), the Global Environmental Management Initiative (GEMI), the Coalition for Environmentally Responsible Economies (CERES), the International Chamber of Commerce (ICC), and the Business Council for Sustainable Development (BCSD). One other group, the Management Institute for

Environment and Business (MEB), is helping business schools to enhance the environmental content of their programs and to prepare future business leaders to compete in the coming age of sustainable development.

Both the *CERES Principles,* issued by the Coalition for Environmentally Responsible Economies, and the *Business Charter for Sustainable Development,* issued by the International Chamber of Commerce, have anticipated the meaning of sustainable performance. The companies that have become signatories to these documents have recognized that the interests of business and the protection of the environment are not in fundamental conflict and that both interests must be integrated into mutually supportive goals. Conflict between the two interests inevitably will result in cataclysmic results for all.

In addition to the CERES and ICC codes, the fifty original executive officers who were members of BCSD have published a *Declaration* in which they describe the requirements for global sustainable development and the new partnerships that are required among businesses and governments.

Sustainable development (SD) is the large-scale, macrodescription of how all nations must proceed in full cooperation with the earth's resources and ecosystems to maintain and improve the general economic conditions of people both now and in the foreseeable future. SD focuses on national and international policy.

Sustainable performance (SP) is the small-scale, microdescription of just what each business and industrial organization must do to translate the concept of sustainable development into practical business applications. SP affirms that if nations are to survive, the businesses of these nations must survive, and that to survive they must profit.

Similarities

The primary similarities between SD and SP stem from the concept of *sustainability*. Both SD and SP are concerned with a future that stretches beyond clearly definable time limits and with the economics of human survival and improvement.

Both SD and SP recognize that many of the current ways in which humans and their business enterprises use the environment are not sustainable. The use of nonrenewable fossil fuels such as coal, oil, and gas at present levels is not sustainable. The current demands made on the earth and its atmosphere as waste sinks are not sustainable. The current levels of toxic emissions are not sustainable. Current agricultural practices of soil destruction through erosion and nonreturn of nutrients are not sustainable. Agribusiness' dependence on nonrenewable energy sources for the manufacture of fertilizers, transportation, freezing, and packaging are not sustainable. The truth that we must face is that neither human life nor the life of much else that inhabits the planet is sustainable if we continue our profligate abuse of the environment.

Within the more general concept of sustainability, there are a number of very specific characteristics shared by SD and SP—although SP gives these characteristics its own special emphasis. I have stated previously that the general characteristics of sustainable development are equity, stewardship, limits, communal, and systemic. These characteristics also are applicable to sustainable performance. However, SP defines these characteristics as they apply to specific organizations, not as they apply to nations.

Equity

SD affirms the position of international and intergenerational equal access to economic well-being. SP affirms the position of interorganizational and intraorganizational equal access to economic improvement. SP promotes the position that each organization must use the environment in ways that permit all other organizations to make use of the environment. SP further supports the belief that an organization has the responsibility to ensure that its future members and stakeholders can compete and profit as well as its present members and stakeholders.

Stewardship

SD affirms that all business processes must be conducted in a way that does minimum harm to the environment. SP affirms the same, but adds that ways must be found to do minimum harm that are financially sound.

Limits

SD affirms that nations must recognize and accept the limits within which the earth's resources and ecosystems can tolerate human use and abuse. SP affirms that organizations must recognize that there are costs associated with the earth's resources and ecosystems that must be included in the organizations' accounting processes and which will place limits on the size and nature of their businesses.

Communal

SD affirms that all environmental problems are now global and that no nation can afford to act unilaterally toward the environment. SP affirms that all business is global and that no business can prosper if other businesses damage the environmental resources on which all depend. For example, when the amount of industrial waste caused by one segment drives up the cost of waste management, all pay; and when another segment exacerbates a pollution problem, all will pay the costs of new, more restrictive legislation. SP requires that all organizations look for opportunities to form links, bridges, and cooperative endeavors among businesses, governments, special-resource groups, and their stakeholders.

Systemic

SD affirms that development must proceed with full awareness of the interlocking relationships among all ecosystems. SP affirms that businesses must not only plan and execute their performance with a full understanding of the interlocking relationships among all ecosystems, but that they must redefine themselves as ecosystems.

SD focuses on the macro issues of global environment, population, poverty, and third-world development. The conduct and future of business will, of course, be influenced by political decisions about all these global issues.

In fact, however, businesses cannot resolve the problem of population—and most often cannot appropriately participate in discussions of the problem. There is no way that the typical business can incorporate population control or noncontrol into its strategic planning.

Businesses cannot devote themselves to the elimination of poverty. They certainly have a stake in the prospering of every segment of the world's economy in order to ensure international markets for their goods and services. Nevertheless, the typical business cannot possibly manage its enterprise in order to eliminate poverty without destroying its own capacity to continue.

No-Regrets Policies and Performance

SD and SP differ primarily in scope. SD is concerned with global economics, and SP is concerned with the business of organizations. However, both support the same response to the current environmental challenge. The response is that nations and businesses enact "no-regrets" policies and immediately undertake no-regrets environmental actions and programs. No-regrets actions are those that make sense for business and for the environment regardless of what the real or final environmental threats turn out to be. In other words, both SD and SP affirm that it makes good sense for businesses to save energy and to find new sources of energy, regardless of how the issues of global warming and climatic change are played out.

The EPA's "Green Lights" program is an example of a no-regrets program to promote energy-efficient lighting installations. The EPA obtained the voluntary cooperation of organizations to replace their current lighting fixtures with more efficient ones. The EPA served as consultant and information clearing house. In addition to saving energy (a prudent business strategy under any circumstances), the program will reduce carbon emissions by 21 million to 55 million tons by the year 2000.

The untried opportunity for energy conservation in the United States has enormous potential financial benefit. The Rocky Mountain Institute, for example, judges that if people in the U.S. were as efficient as the Japanese in the utilization of energy, they could save $200 billion per year [7].

The advanced energy technology being produced in Japan will have increasing competitive significance. In the field of solar energy, for example, Kansai Electric Power and the Sharp Company have produced a solar heating and air conditioning unit that is integrated with traditional electrical energy and which can heat and cool a two-story house [3].

Reducing the use of energy; reducing waste; developing drought-resistant crops; developing automobiles with greater fuel efficiency; and installing heating, ventilation, and air-conditioning controls that maximize efficient energy use all make monumentally good business sense, regardless of their impact on the environment. In fact, they all have very positive results for the environment.

Characteristics of Sustainable Performance

Sustainable performance has two fundamental characteristics that distinguish it from sustainable development. These are profit and performance.

Profit

Profit is not a key element in SD but it is in SP. However, even though profit is not addressed in descriptions of SD, everything that has been written and said about SD assumes that business can improve the per-capita income of the world's population at some optimum rate that is not damaging to the environment. Whereas SD makes the implicit assumption that it is the profit of business and industry that will deliver the growth in per-capita real wealth to ensure "development," SP makes this assumption about profit explicit and central.

Improved profit can occur though cost savings, cost avoidance, new products and services, and increases in the selling price of a service or product. The EPA has identified a large list of cost-saving and cost-avoidance opportunities available to organizations just by reducing waste [6]. The list includes the following opportunities:

- Reduced waste-management costs
- Cost savings for input material
- Insurance and liability savings
- Changes in costs associated with quality
- Changes in utilities costs
- Changes in operating and maintenance labor, burden, and benefits
- Changes in operating and maintenance supplies
- Changes in revenues from production
- Increased revenues from by-products

One very important difference between the concepts of SD and SP is that SP makes an explicit commitment to the concept of profit as a key driver toward environmental performance. Another key difference is that SP emphasizes performance and not development.

Performance

The emphasis in SD is development. The emphasis in SP is performance. Performance includes all the actions that individuals and teams undertake to perform the work of an organization and to achieve the goals that sustain the organization. The idea of performance does not necessarily include certain meanings associated with development.

Development carries with it the ideas of growing, increasing, expanding, or enlarging. The key qualifying idea for performance is none of these. The key qualifying idea for performance is *improving quality.*

We may agree that the responsibilities of nations and governments include increasing the wealth and raising the standards of living of their people. We may agree that these responsibilities include improving the economic conditions of their poor. At present, however, there are such links among the factors of environment, poverty, trade, debt, and consumption that it is hard to envision an achievable solution to the problem of poverty. The monumental debt burden of developing countries leads them to deplete their forests, fisheries, and nonrenewable resources to increase their earnings from export. It leads them to make environmental protection a low priority and to transfer responsibility for global environmental renewal to the rich countries. Until the problem of debt is resolved, the problem of poverty will not be resolved [1,2,5].

Poverty is an evil that should command the full attention of government. The facts are, however, that governments have not resolved the problem of poverty (in spite of a host of policy changes and programs), and both the absolute and relative numbers of the poor are still increasing. In Africa alone, the number of poor people will rise by about 50 percent by the year 2000 [13].

Sustainable development assumes that development can continue indefinitely at some sustainable pace and that all the earth's peoples can enjoy a standard of living similar to that of the rich countries today. This is an act of faith of mythic dimensions. The most optimistic analysis of the earth's recoverable mineral and energy resources shows that there is no chance of all the five billion people who are living today rising to such a standard—much less that the ten billion projected for the next century will [16]. For SD to make any sense, its goal must be defined in terms other than that everyone on earth today or tomorrow is going to rise to the level of wealth enjoyed by today's production-consumer societies.

As an economic and political strategy for responding to the problem of the earth's environment, SD probably will have no greater success in taking care of the poor and leading to an equitable distribution of wealth than have any of the past political and economic strategies. The fallacy in all such strategies is the assumption that development will always be possible, that development leads to increased national wealth, and that such wealth will solve the problem of poverty [13].

SP affirms that the first goal of business is not to find ways to grow and expand. The first goal of business is not development. The first goal is total quality and the continuous improvement of the processes, services, and products of business that are required in the environmental age. It is only by serving this goal first that the goals of environmental enhancement, long-term profit, and competitive position can be maintained. *Quality is the real key to sustainability in business—not development.*

SP can be achieved without growing or expanding anything except profit. SP inevitably will lead organizations to use less energy and material resources. It inevitably will lead industry to produce fewer products with longer life cycles. In pursuing SP, organizations will discover green market niches for new services and products to replace their current, environmentally unsound ones.

The only increases that organizations should be concerned with are, first, the increase of quality, and, second, the consequent increase of profit. Profit, as I have noted above, is not necessarily a function of more. It can be a function of less—less use of energy, less use of raw materials, less creation of waste, less use of time, and so on. SP in the environmental age will lead organizations first toward the continuous improvement of their services and products so that they are more compatible with the earth's ecological systems. The pursuit of this goal will lead them inevitably toward finding new opportunities for profit.

Profit and performance are two key characteristics of SP. SP is, however, more complex than a set of characteristics.

SP represents a new way of understanding the organization as a system. It radically redefines the traditional relationships among the system elements of input, work process (transformations), and output.

A Systems Model for Sustainable Performance

The traditional open-systems model of organizations includes three elements: input, work processes (transformations), and output. Expanded models may include suppliers and customers. None of the traditional models include the environment, but it is the environment that now must be taken into account and that, along with the traditional elements of an open-systems model, provides the basis for understanding SP.

Figure 4-1, the Systems Model for Sustainable Performance, shows the environment as the context within which organizations must plan and execute performance. The model shows the environment interacting with the three traditional elements of input, process, and output. It also suggests that the environment creates a new dimension for the ways in which suppliers and customers are viewed and managed.

132 / Competitive & Green

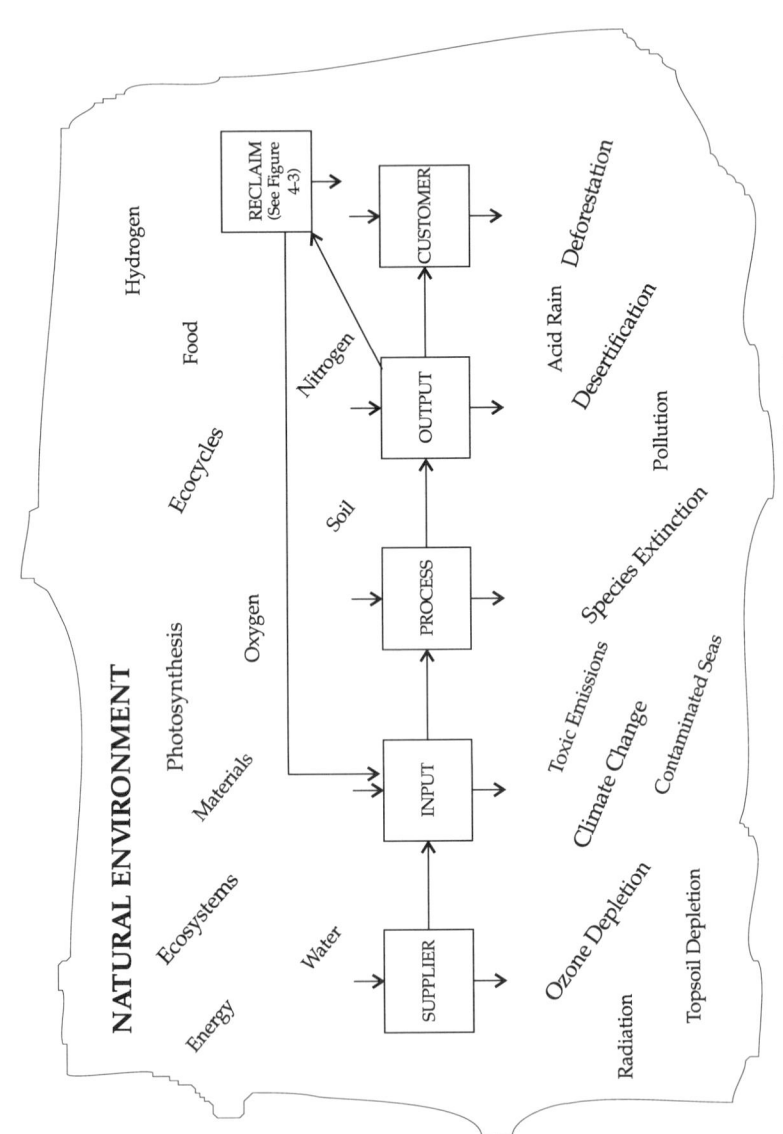

Figure 4-1. Systems Model for Sustainable Performance

The model contains the following elements:

- Natural environment
- Supplier
- Input
- Process
- Output
- Customer

The model also contains the element of flow, which is indicated by arrows. Materials, energy, air, and water flow into the organization. The traditional output of products and services flows to the customer. In addition, potential wastes, toxic emissions, and pollutants flow to the customer and eventually flow into the environment. Wastes, toxic emissions, and pollutants also flow directly into the environment from the organization's processes, products, and services.

One way to explain the model and SP is that the environment must be perceived as a sole-source *supplier* that must be paid and kept because there is no other supplier available; also, the environment is a *customer* who is the final arbiter of quality. Each element in the model is discussed below, with specific implications for SP.

Natural Environment

There are two ecosystems that, taken together in their multitude of interactions, form the life system on which all humans depend. The first is the natural system, composed of the vast network of life-creating and life-supporting activities of countless microbes, plants, and other organisms on a scale that is much too large to understand fully. The second system is the agricultural system created by humans.

The element of environment in the Systems Model for Sustainable Performance refers to the natural ecosystem.

This enormous ecosystem and its multitude of integrated cycles and subsystems support human life, business, and society by

1. Maintaining the gaseous composition of the earth's atmosphere.
2. Keeping climate changes sufficiently gradual so that life forms can adapt to such changes.
3. Regulating the earth's water cycle.
4. Generating and maintaining soils.
5. Disposing of wastes and cycling nutrients.
6. Maintaining the global carbon cycle and nitrogen cycle.

For a very long time, the ecosystems of nature were so large and so self-regulating and so powerful that the impact of human beings on these cycles and systems resulted in only minor perturbations. Nature was always capable of coming back into equilibrium, with little the worse for being a bit out of kilter. Unfortunately, now the relation of human beings to the environment is altogether different. Human enterprise has triumphed, and nature no longer exists as an independent force.

This book is not about ecology, ecosystems, and ecocycles. It does not attempt to describe the very complex organization of nature and the place of humans in this organization. However, even without a deep understanding of ecology, most people are able to see that we cannot continue to treat the environment as a free dump with an endless capacity to absorb the wastes of production and consumption.

End-of-pipe management of wastes, toxins, and pollutants was a workable strategy so long as the amounts of discharge were a fraction of nature's potential for managing those discharges and so long as there appeared to be no limit to the capacity of nature's waste sink. We are fast approaching that limit; indeed, it may have been reached. The self-adjusting capacity of the earth's ecosystems has been severely impaired or destroyed.

It must be realized that every element in an organization's performance interacts with the environment. The Systems Model for Sustainable Performance depicts each element as receiving inputs from the environment and discharging outputs into the environment. Input from the environment and output into the environment can no longer be viewed as occurrences at the traditional boundaries of an organization's system—at the interface with suppliers and customers. This cycle of input from the environment and output into the environment must be understood in terms of the full history of every input, that is, how environmentally friendly are the supplier's processes and products and those of its suppliers? The cycle of input and output exists in every step of every process employed by an organization. This cycle operates with each output from the organization and with the organization's customers and the customers' customers, and so on, until the product is recycled, reused, or treated as waste. It is this very radical understanding of the environment's premier place in the system of an organization that is the heart of sustainable performance.

Supplier

The next element in the model is the supplier. Suppliers, like all other elements, now have a new, ecological relationship to organizations. They no longer can be considered just as sources of energy, raw materials, product components, capital, and a host of special services such as maintenance, repair, and expertise. They must be considered from the perspective of their ecological impact. Suppliers no longer can be selected just because they provide input that is fit to use 100 percent of the time—unless, of course, we redefine "fit to use" in terms of the environment's ability to use it.

Suppliers are elements in every organization's ecological system and must be evaluated, not only for the quality and timeliness of their inputs, but also for how they impact the environment through their own processes of production and the specific environmental impact of the services and products they deliver.

In SP, suppliers are taken into the organization's own performance system and are expected to function in ways that are fully congruent with the values and purposes of the organization. It is not possible for an organization to manage the full life-cycle of its own products without knowing the history that the raw materials, components, and services have had with its suppliers.

Organizations that endorse SP for themselves must require it of their suppliers. This means that suppliers will be expected to

- Demonstrate their commitment to the principles of sustainable performance (discussed in the next chapter).
- Have in place an active program to measure their own environmental performance (one such as the Sustainable-Performance Assessment discussed in Chapter 7).
- Use tools such as environmental audits, process improvement, measurement, and life-cycle planning on a routine basis.
- Demonstrate specific improvements (planned and actual) in their environmental performance.

Input

The model shows that there are two kinds of input into an organization's work processes. One kind of input comes directly from the environment, and the other comes from suppliers.

Environmental Inputs. The model communicates the idea that input from the environment must be recognized explicitly as input. The current shape of the environment has been created largely by consumers' and suppliers' not recognizing the critical importance of the input from the earth's ecosystems into the systems of production. Up to now, decisions about performance have been based on a simplistic economic model that has been indifferent to nature's obvious, negative-feedback signals of ozone depletion, deforestation, desertification, and loss of wetlands. We have persisted in using an unrealistic accounting process in assessing the cost of doing business.

In business, inputs that are not acknowledged are treated as free and are not included in accounting records or balance sheets. The unacknowledged inputs from nature have not been costed, no tax has been charged for their use, and they have been viewed as contributing nothing to the costs of doing business.

One major weakness in the economic and accounting models that we have used is that they fail to take into account such things as the many subsidies that perpetuate enterprises that are indifferent to the environment. Subsidies to the energy industry in the U.S. amount to hundreds of billions of dollars per year, if one includes tax credits, costs to the environment, health-care expenditures, and the like [1].

It is beyond the scope of this book to discuss in detail how environmentmental inputs can be valued. Many different approaches currently are being discussed [11]. If it occurs, "costing" of input from the environment probably will be in the form of a tax. The tax will be based on a projected computation of environmental impact. This impact will include consideration for depleting a resource or for producing pollution, toxicity, and/or waste.

The U.S. Congressional Budget Office has determined that a tax on fossil fuels of $28 per ton of carbon content would not only discourage future increases in carbon-dioxide emissions but also would raise $163 billion for the U.S. Treasury. It is estimated that a tax of $113 per ton of carbon content would reduce emissions by up to 20 percent and would generate $190 billion in revenue. Taxes such as this could be applied in a fiscally neutral way by incentives and offsetting tax reductions. For example, organizations could be given bonuses for exceeding EPA requirements controlling the generation of wastes, toxics, and pollutants. They could be given investment credits for upgrading plants and machines in ways that improve environmental performance, and corporate tax rates could be reduced by amounts consonant with taxes against the carbon content in fuel.

One factor that contributes to our failure to place a cost on commerce's use of the environment is that the economic indicators that we typically use fail to take the environment into account. We track and report economic trends such as consumer confidence, consumer price index, retail sales, gross national product (GNP), home sales, and the like—business and economic changes such as rates of expansion and contraction. We do not track and report as economic news environmental data on loss of topsoil, loss of fish harvest, growth in air pollution, rate of desertification, old-growth forest loss, energy use, and the like. What is needed is ecological *and* business reporting.

It must be emphasized that input from the environment must be valued, and the most universally understood way to value it is to put a price on it. Air, land, water, and nonrenewable resources must be valued in relation to sustaining the natural environment. The many services provided by the environment are not free.

Suppliers and Inputs. Total quality management has taught business that it must invoke the highest possible quality standards on the input that it receives from suppliers. "Just-in-time" supply and inventory systems and "partnering" with suppliers are popular initiatives that developed from the need for organizations to impose total-quality standards on the inputs received from their suppliers. It is clear that an organization can be world class only if it has world-class suppliers.

SP involves the same concern for quality of inputs. It just extends the meaning of quality to include environmental quality. In assessing the environmental quality of the input that suppliers deliver, organizations will ask the following kinds of questions:

- What solid wastes will be produced, and what will be the costs for managing them?
- What toxic wastes will be produced, and what will be the costs for managing them?
- What emissions will be produced, and what will it cost to manage them?
- What will be the risks and costs associated with handling and storing any materials, chemicals, or products that are delivered to us?
- What is the recyclable potential of the input?
- How easily can recyclable materials be separated from other materials at the end of a process or in end products?

- What is the weight per volume of products and packaging? Can it be reduced?
- How much recycled material did the supplier use?
- How repairable is the product or element received?
- What is the life span of the product received compared to known benchmarks for such products?
- How much packaging is used? Can it be reduced?

Including "environment" as the context within which the system of input, process, and output occurs fundamentally changes the way in which we understand organizations.

Process

"Process" and "work process" are the terms used throughout this book to describe any repetitive sequence by which any work is accomplished, any service delivered, and any product produced. A process is the flow of an object through a sequence of steps that includes transport, delay, operations, and inspection [8]. An object is whatever is moved or changed. It may be a report, idea, metal stock, chemical compound, or item such as an automobile frame. Figure 4-2 shows two examples of the work-flow process.

Simplified Manufacturing Process

INPUT						OUTPUT
STEP	Delay	Transport	Operation →	Transport	Delay	Operation →	Transport	etc.
DESCRIPTION	Storage	Move raw materials to machine	Cut	Move to next machine	Wait in line	De-burr	Move to storage	

Process ──────────→

Simplified Engineering Process

INPUT						OUTPUT
STEP	Operation →	Transport	Operation →	Transport	Delay	Inspect	Transport	etc.
DESCRIPTION	Drawing request	Deliver request to drafter	Prepare drawing	Deliver drawing to engineer	Drawing waiting for engineer	Verify drawing	Deliver to printing	

Process ──────────→

Figure 4-2. Work-Flow Process

The quality of processes in the Systems Model for Sustainable Performance, like all the other elements, is determined by imposing traditional criteria such as efficiency and reliability, but these processes now also must be evaluated for their impact on the environment. Each step in a process typically exchanges something with the environment and has an impact on the environment. Figure 4-3 displays the inputs and outputs of a process that helps to identify the various opportunities for continuous environmental improvement. This figure also is useful as a tool for evaluating the environmental impact of new processes as they are developed.

The points to be emphasized are as follows:

1. The quality of a process can be improved by eliminating or improving any of the steps involved, i.e., transportation, delay, operation, and inspection.
2. The quality of a process also must be improved in the way in which it exchanges energy and materials with the natural environment.

Any process provides an immediate opportunity for improvement for sustainability. A 3M facility, for example, changed from the use of an acid solution to the use of a slurry of water and pumice to clean copper sheeting for making circuit boards. The change eliminated 40,000 gallons per year of hazardous waste as well as the cost for disposing of the waste [11].

We have now examined environment, suppliers, input, and processes. The next element in the Systems Model for Sustainable Performance is output.

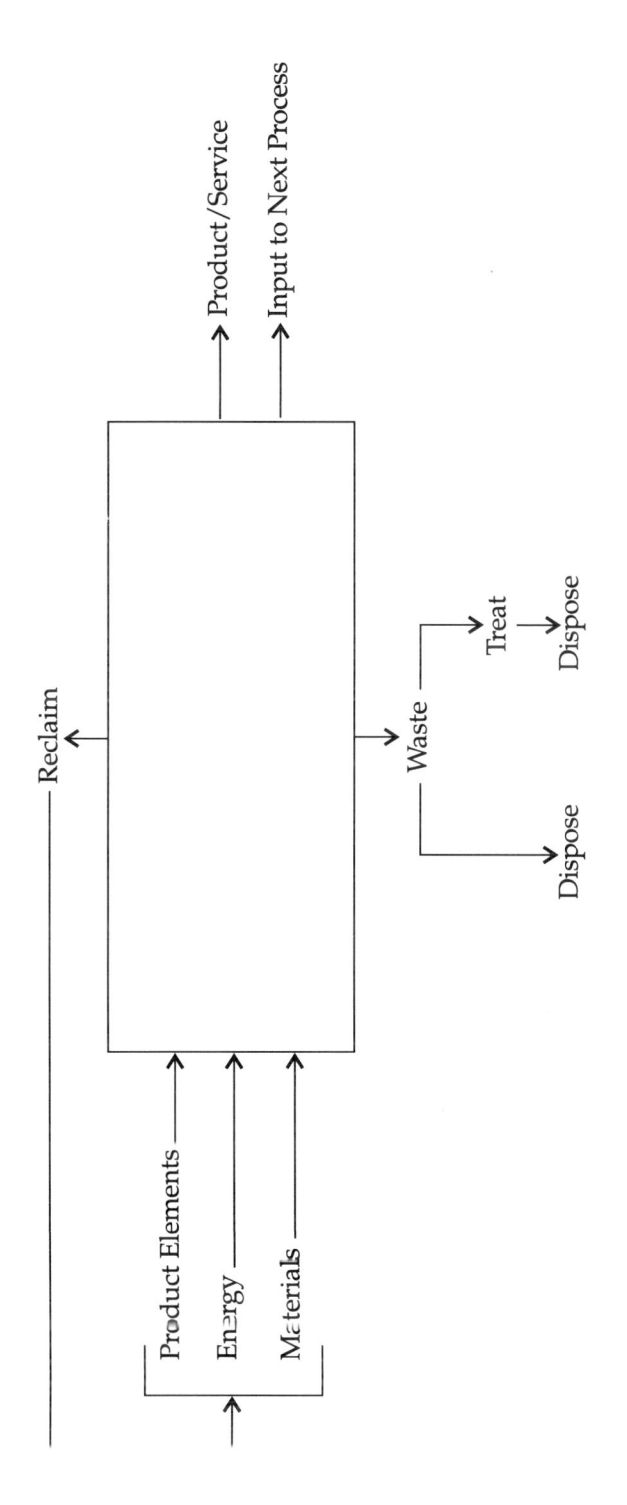

Figure 4-3. Process-Improvement Opportunities

Output

Waste and pollution that is not managed during input or processing must be managed at the point of output. The most desirable way to reduce waste is to prevent it prior to its becoming output. If waste is managed only by "end-of-pipe" or "out-the-back-door" techniques, the waste is simply transferred from one medium or place to the next. For example, conventional water treatment removes waste as sludge and then puts the sludge in a landfill, which removes it from one medium and place and shifts it to another medium and another place.

Consideration for the environment greatly complicates the way in which output must be understood and managed. The following are types of output:

- Traditional outputs of *services and products* to customers.
- Outputs that become *reclaimed waste* and that are recycled into a process or sold to external customers.
- Outputs that are unrecovered waste (solid, hazardous, toxic emissions, pollutants) that must be treated and/or deposited somewhere. Unrecovered waste is *direct output to the environment.*

Outputs (Services and Products). For outputs of services and products to customers to be sustainable they must carry with them the potential for doing minimal harm to the environment throughout the full cycle of their use. SP forces a company to include its customers, as well as its suppliers, in the life-cycle management of its products and services. This means that an organization will ask of itself and its outputs the same questions that it will put to its suppliers.

- What solid wastes will our outputs produce and what will be the costs to our customers for managing them?
- What toxic wastes will our outputs produce and what will be the costs to our customers for managing them?
- What emissions will be produced by using our outputs and what will be the cost to our customers for managing them?
- What will be the risks and costs to our customers associated with handling and storing any materials, chemicals, or products that we deliver to them?
- What is the recyclable potential of the output?
- Can recyclable materials be separated from our inputs?
- What is the weight by volume of products and packaging? Can we reduce it?
- How much recycled material was used in the output of the product?
- How repairable is the product or element?
- What is the life span of the product we have delivered compared to known benchmarks?
- How much packaging have we used? Can it be reduced?

Evaluating packaging is one example of considering output from the point of view of creating waste. In evaluating the environmental impact of the way a product is packaged for a customer, the following ways of reducing waste should be considered:

1. Packaging that is not independent of the product itself but can be used as part of the product and is consumed with the product.
2. Minimal packaging.
3. Packaging that can be reclaimed and reused (e.g., glass bottles).
4. Packaging that can be reclaimed and recycled (e.g., paper, plastic, and wood).

An example of innovative packaging that is consumed with the product was developed by DuPont for marketing in India. A Canadian subsidiary of DuPont developed a polyethylene pouch container for milk. These pouches reduce solid waste by up to 70 percent, compared with traditional ways of packaging milk. This technology was adapted to India's need for a way to dispense cooking oil. The low-cost pouch that DuPont introduced holds the measure for cooking a typical meal, and the polyethylene film itself dissolves at cooking temperature and is edible. The result is zero waste caused by packaging [14].

Reclaimed Waste. Organizations not only deliver output to customers (and, indirectly, post-customer waste into the environment), they also produce waste that is either reclaimed and recycled back into a process, used as input in another process of the organization, or sold to an external customer. Reclaiming waste often involves separating waste streams to reduce the volume of hazardous waste that must be handled or to facilitate the reuse of waste in a process.

Waste recovery may, of course, not occur just at the end of a process. It may occur at any point during a process and further downstream, after the use of an output by a customer or at some post-consumer point.

The General Electric Company (GE) in Wilmington, North Carolina, has achieved considerable success in recovering its waste and turning waste into profit. It now recovers 800 tons of ammonia annually with equipment that will permit recovery of investment in five years. The company also has implemented a centralized trash collection and sorting operation to recover 20,000 cubic feet of reusable and 50,000 cubic feet of recyclable materials per year. Through these and other recovery projects, GE has achieved a savings of $3.73 million per year.

One inventive new way to package is by using molded fiber technology (MFT). The products resemble egg cartons and are made of water and recycled paper. One cosmetic manufacturer who uses the material reports that it has reduced the "crumble in transit rate" from 34 percent to 10 percent. The material is customized to fit each packing need and has proven effective as packaging for computer elements.

Printers now use ink-recycling units to produce black newspaper ink from various waste inks. These units blend the different colors of waste ink together with fresh black ink and black toner to create black ink. The ink is used in place of fresh black ink and eliminates the need to ship waste ink offsite for disposal.

The Rexham Corporation installed a distillation unit to reclaim n-propyl alcohol from waste solvent. The installation cost was $16,000. The company now recovers 85 percent of the solvent in the waste stream, which results in a yearly savings of $15,000 in virgin-solvent costs and $22,800 in waste-disposal costs [10].

Output may go directly to customers in the form of services or products and it may be reclaimed to be recycled or reused. What is not delivered to a customer or not recovered becomes output into the environment.

Direct Outputs to the Environment. Outputs that do not go to the customer, or which are not reclaimed, go into the environment either before or after treatment. Direct outputs to the environment exist along a scale that runs from very useful to very harmful. Our concern is with those that are harmful and which should be minimized.

Wastes are harmful by virtue of the ways in which they interact with the environment. Harm is created by the nature of a waste (its toxicity, nondegradeabilty, or radioactivity) as well as its volume. For example, certain amounts of the discharge of CO_2 by automobiles, airplanes, trains, trucks, and power plants can be tolerated by nature. However, after a certain level is reached, it threatens to cause global warming and rapid climatic change.

Serious problems at the point of managing harmful output into the environment should force an organization to recognize that it is not managing pollution and waste "upstream" at the point of supplier input and work processes. The preferred strategy for managing harmful output into the environment always is to increase efforts at prevention.

Managing harmful output into the environment involves some sort of transfer technology, i.e., transferring some portion of the waste or pollutant to some other place or media. The most common methods for doing this are composting, incineration or combustion, chemical and mechanical treatment, and landfills. None of these are satisfactory, and landfills are becoming less and less of an option.

The EPA reports that the amount of solid waste generated in the U.S. doubled from 88 million tons in 1960 to 180 million tons in 1988. By the year 2000, the U.S. is expected to generate 216 million tons of solid waste per year—and this figure takes into account a very optimistic projection for recycling. More than half the people in the United States live in areas with fewer than ten years of landfill capacity left. By 1995, all landfills in New York State will have reached capacity and will be closed [10].

Incineration of hazardous waste results in the discharge of very toxic chemicals and produces about 400 million pounds of toxic ash each year. The EPA indicates that 1.4 billion pounds of hazardous wastes were incinerated in 1988 and that this amount will double by 1993. Incineration is not a solution to pollution and waste. Like all other end-of-pipe treatment technologies, it solves one problem and creates others.

Composting is not much of an industrial alternative in the U.S., although it has been growing quite rapidly in Europe. In Europe, there are over 200 facilities in use; in the U.S., there are about a dozen. Proctor & Gamble has taken the lead in forming the Solid Waste Composting Council, which has a goal of popularizing composting as a part of municipal solid-waste disposal systems.

We have now looked at all the elements in the Systems Model for Sustainable Performance except the customer. The relation of an organization to its customers, insofar as the environment is concerned, consists largely in thinking about the customer as an extension of the organization's own system of input, work processes, and output.

Customer

Customers interact with the environment, as do all other elements in the model. The goal in SP is to deliver to the customer services and products that create the minimum number of environmental problems. The proper way to read the Systems Model for Sustainable Performance is to start with the customer.

The services and products delivered should be responsive to the customer's needs. The needs of a customer can never be determined in theory. They must be determined through direct contact and exploration. A preliminary assessment of a customer's needs, however, can be done by asking of the organization the same sort of questions that it asks of its suppliers.

- Questions about the wastes that a service or product will produce and how much it will cost the customers to manage these wastes.
- Questions about the risks and costs for the customers associated with handling and storing any materials, chemicals, or products that are delivered to them.
- Questions about the recyclable potential of the materials and products that the organization provides its customers.
- Questions about the life cycle and repairability of what the organization delivers.
- Questions about volume and weight per unit of products and packaging and questions about the amount of packaging.

Sustainable performance requires more than delivering a service and product that causes minimum problems for customers. SP requires that the organization educate the customer in the best and safest use of all its products. It means sharing with the customer new environmental technology and helping customers to establish their own programs for SP.

Summary

Although sustainable performance (SP) is a logical and historical derivative of sustainable development (SD), it differs from SD in a number of important ways. SD is the large-scale, macrodescription of how all nations must proceed in full cooperation with the earth's resources and ecosystems to maintain and improve the general economic conditions of people both now and in the foreseeable future. SD focuses on national and international policy. SP is the small-scale, microdescription of just what each business and industrial organization must do to translate the general concept of SD into practical business applications. SP affirms that, if the nations of the earth are to survive, the businesses of these nations must survive, and that to survive they must profit.

One key to understanding SP is the Systems Model for Sustainable Performance, which adds the natural environment as the primary element to the traditional open-systems model of organizations. This chapter has described the elements in the Systems Model for Sustainable Performance and has suggested the implications that these elements and their relationships have for doing business in the environmental age. It also has described two dominant characteristics of SP, profit and performance, and has provided examples of these characteristics from the environmental initiatives of organizations.

However, SP represents a more far-reaching and thorough change than I have thus far described. It represents a fundamental shift in the underlying values or principles that guide the way organizations define themselves. In the next chapter, I will list these principles and describe their implications for planning and implementing SP.

References

1. Barbier, E. (1989). *Economics, Natural Resources, Scarcity and Development.* London: Earthscan.
2. Barbier, E. (1987). The Concept of Sustainable Development. *Environmental Conservation, 14*(2), 101-110.
3. *Business Week.* (July 15, 1991). p. 131.
4. Elkington, J., & Knight, P., with Hailes, J. (1991). *The Green Business Guide.* London: Victor Gollancz, Ltd.
5. Elkins, P., Hillman, M., & Hutchison, R. (1992). *The Gaia Atlas of Green Economics.* New York: Anchor Books.
6. Environmental Protection Agency. (1990). *Guides to Pollution Prevention.* Washington, D.C.: Author.
7. Henderson, H. (1991). *Paradigms in Progress.* Indianapolis, IN: Knowledge Systems.
8. Kinlaw, D. (1992). *Continuous Improvement and Measurement for Total Quality: A Team-Based Approach.* San Diego, CA: Pfeiffer & Company.
9. Pearce, D., Markandya, A., & Barbier, E. (1989). *Blueprint for a Green Economy.* London: Earthscan Publications Limited.
10. Price, R. (April, 1990). Stopping Waste at the Source. *Civil Engineering,* pp. 8-10.
11. Smart, B. (1992). *Beyond Compliance.* Washington, D.C.: World Resources Institute.
12. Tolba, M. (1987). *Sustainable Development: Constraints and Opportunities.* London: Butterworth.
13. Trainer, T. (1989). *Developed to Death.* London: Meadows, Meadows & Randers.

14. Woolard, E. (Spring, 1992). An Industry Approach to Sustainable Development. *Issues in Science and Technology*, pp. 29-33.
15. World Commission on Environment and Development. (1987). *Our Common Future.* New York: Oxford University Press.
16. World Bank. (1992). *World Development Report 1992.* London: Oxford University Press.

5

THE PRINCIPLES OF SUSTAINABLE PERFORMANCE

Sustainable performance is based on a new way of viewing the traditional elements of input, process, and output in a general systems model. SP has distinctive characteristics that define the limits within which performance must be conducted in the new age of the environment. SP also represents a major change in the underlying values or principles for doing business.

TQM involves a similar shift in values. Leaders who have implemented TQM have had to shift away from

- Defining quality by finite standards, to defining quality by customer perceptions and expectations.
- Focusing on "putting out fires" and "fixing what's broken," to focusing on systemic change and finding new ways of doing business.
- Controlling people, to helping them find higher and higher levels of personal and team influence.
- Communicating ideas to people, to stimulating them to develop their own ideas.
- Focusing on individuals as primary work units, to focusing on teams as primary work units.
- Concentrating on functional responsibilities, to concentrating on process and system responsibilities.

- Encouraging internal competition, to encouraging internal cooperation.
- Telling people what they need to improve, to helping them search for and implement their own improvements [9].

SP assumes these TQM shifts, but adds the major new dimension of shifting away from performance that *uses* the environment to performance that *cooperates* with the environment. This shift of values can be incorporated into a set of principles. These principles, taken together with (1) the short definition of SP; (2) the description of SP's similarities to SD; (3) the description of SP's two distinguishing characteristics of profit and performance; and (4) the Systems Model for Sustainable Performance provide a comprehensive description of the meaning of SP.

The purpose of this chapter is to describe these principles and to show how they can be used as further guidelines for planning and implementing SP.

Principles of Sustainable Performance

SP is more than an adjustment in the way that organizations have been doing business. It represents a radical and fundamental shift in orientation and values [12]. For organizations that have become fully committed to total quality management and continuous improvement, this shift will be considerably less difficult than it will be for those organizations that have yet to assume the values of total quality, zero error, team development, and total customer satisfaction.

SP represents a new way of understanding the organization and the quality of leadership required in the environmental age. This new understanding is expressed in the following set of principles.

Principle One. Sustainable performance is a process of systems thinking, analysis, and integration that requires that the organization be understood and managed as a system.

Principle Two. Sustainable performance is an ecologically interdependent process and requires that all organizational processes, products, and services be revised or replaced to ensure their compatibility with nature's ecosystems [10].

Principle Three. Sustainable performance is a results-oriented process and requires the demonstrated commitment of organizational leaders to specific, measurable results.

Principle Four. Sustainable performance is a community-building process. This requires organizations to cooperate with one another and use the environment in ways that are equitable for one another. This also requires that organizations involve all their stakeholders in the processes of planning and implementing sustainable performance.

Principle Five. Sustainable performance is a limiting process. It requires that organizations recognize that there are costs associated with the earth's resources and ecosystems that must be included in the organizations' accounting processes and which will place limits on the size and nature of their businesses.

Principle Six. Sustainable performance is an open process and requires that organizations communicate fully all aspects of their planned and actual environmental performance to all organizational stakeholders [1]

Principle Seven. Sustainable performance is a process of continuous improvement of every aspect of an organization's performance and requires the full involvement of every member of the work force [7,8].

Principle Eight. Sustainable performance is a data-based process and requires concrete information retrieved from auditing, measuring, and reporting the organization's environmental performance [1].

Principle Nine. Sustainable performance is a technologically dependent process and requires organizations to develop partnerships with governments, other organizations, educational entities, research and development sources, suppliers, and customers in order to discover and implement ways to improve sustainable performance [2,3].

Principle Ten. Sustainable performance is a total organizational process and requires that all planning, decision making, and human resource systems be made fully congruent with the organization's commitment to sustainable performance.

Uses of the Principles

The Principles of Sustainable Performance have the following applications:

1. To help define the meaning of SP and provide content for communicating this meaning.
2. To serve as guidelines for planning the steps to SP.
3. To provide a device for assessing an organization's progress toward SP.
4. To provide guidance for selecting and employing specific strategies and tools for SP.

Principle One

Sustainable performance is a process of systems thinking, analysis, and integration that requires that the organization be understood and managed as an open system.

This first principle emphasizes the points already made in the descriptions of the Sustainable-Performance Management Model in Chapter 1 and the Systems Model for Sustainable Performance in Chapter 4. Every business and industrial organization is an ecological system whose inputs and outputs affect the earth's total ecological system. To understand this concept and to act on it, leaders must think of their organizations as systems and must employ systems thinking in the ways in which they analyze problems and make decisions.

The problems that we are having with the environment are extensions of a more general problem that has been a persistent nemesis for organizations and their managers. This problem appears in many forms and has been described in many ways, e.g., suboptimization, short-term perspective, destructive competition, poor resource allocation and use, and functional fragmentation. In brief, the problem is that leaders of organizations have persisted in thinking in a compartmentalized, fragmented, or dissociative way. They have not learned to think systemically and have, therefore, not employed the tools of such thinking in their management strategies and practices.

Managing from a systems perspective means a number of things and requires many fundamental changes in the ways leaders think and act. Two changes are especially relevant to SP:

- Organizations no longer can be viewed in terms of parts or functions. They cannot be understood by looking at the properties of each subelement. An organization is not the sum of its parts. It can be understood properly only in terms of interdependent relationships and interactions. It must be viewed holistically and managed by organizing principles and processes.

- Organizations must be viewed as self-adjusting organisms that are controlled by a complex network of interconnected information channels. These channels provide feedback to the system and help it to adjust and remain stable. These channels also may provide information that does not aid adjustment but leads to undesirable results and even the breakdown of the system.

Principles and Processes. The first implication of the principle of systems thinking for SP is that SP is a function of how well the organizing principles and processes of the organization are managed. It requires that leaders struggle continually to see more than the parts and components of their organization. It requires them to understand the underlying principles or values that organize the many different functions of the organization. It requires that they see work processes or flows—rather than functions or responsibilities—as the key elements in performance.

The Principles of Sustainable Performance provide a way to begin thinking of the organization as a system. These principles may not be exactly the ones that will work in every organization in every circumstance. Their major worth may be to encourage systems thinking and acting.

The worst thing that any leader can do is to begin to act and force actions within an organization without having first understood the existing organizing values or principles of the organization. The degree of readiness that an organization has to embrace SP often can be inferred from the level of response that the organization is currently making to the environmental challenge. The next chapter contains an analysis of these response levels.

For SP to develop within an organization, its organizing principles must become fully congruent with SP. I have been actively involved in recent years in assisting a variety of organizations to plan and implement TQM. I have had the opportunity to observe the planning and implementing process in many more organizations. Over and over again, I have seen the same kinds of errors made. Probably the most frequently recurring error is that leaders begin to act before they understand the current organizing principles of their organizations. They import programs from other organizations that will waste time and energy and generate unproductive instability because these programs are based on a different set of organizing principles than those currently operating in the organization.

Here are a few recent examples:

- A closely held company began an extensive team-training program without addressing the fact that the company was a paternal dictatorship with a rigid, hierarchical control system.
- Another company tried to develop a "team culture" without changing its human resource management systems, which encouraged competition and rewarded individuals.
- The head of an engineering-research organization decreed that all major subunits within the organization would develop performance measures at a time when measures were viewed as "bean counting" or ways to threaten and punish people.

Systems thinking means looking for organizing principles. In relation to SP, it requires that leaders address as a first priority the challenge of developing principles within their organizational systems that are fully congruent with SP.

A second requirement of systems thinking is to look for the processes that connect functions and operations. Management by processes is one of the important changes that TQM has brought to management practices. This change has helped managers to look beyond results to the ways that results are produced. Managing for SP requires that we look at the processes by which waste and pollution are produced rather than looking at how to manage the waste and the pollution. A direct application of systems thinking in SP is life-cycle analysis.

Life-cycle analysis (LCA) directs attention to the entire process by which a product or service is created and delivered. LCA is a systems way of thinking. The traditional practice of end-of-pipe waste management is a nonsystems way of thinking. End-of-pipe thinking is the sort of thinking by which managers have tried to manage performance by focusing on results or by changing quality by inspecting it into a product or service—rather than looking at the processes by which results and quality are achieved.

One of the major hurdles to be scaled in the change to SP in organizations is the inability of leaders to use systems thinking and to manage through processes. The failure to think and act systemically is evident in the simplest processes of interpersonal communication. I have analyzed the behaviors of managers in one-to-one problem-solving conversations in the laboratory, in training sessions, and in real-life settings. Without rigorous training and feedback, the majority of managers cannot approach a conversation as a system in which all the behaviors of themselves and others are in continual interaction and adjustment. Instead, they focus on single bits of conversation, single skills, and single variables, without placing their own behavior within the communication system that they have created [6].

A great many books have been written about the time wasted in meetings and how ineffective and inefficient most meetings are. Here again, the problem is not any one event or any one variable. The problem cannot be fixed by doing one or two things such as setting an agenda, sticking to a schedule, or improving communication skills. The meeting must be viewed as a process, and it can be improved only by managing all the interacting variables of this process [9].

The first characteristic of systems thinking is to view the organization as more than its parts or functions. Systems thinking for SP requires that leaders look at the organization holistically and manage in terms of its organizing principles and processes. It requires that leaders help the organization to develop values that are fully congruent with SP. It also requires them to pay attention to the processes by which waste and pollution are produced rather than to manage their disposition.

Connectedness. Another implication of systems thinking is that leaders recognize that their organizations are self-adjusting organisms controlled by a complex network of interconnected information channels. These channels provide feedback to the system and help it to adjust and remain stable. These channels also may provide information that does not aid adjustment but leads to undesirable results and even the breakdown of the system.

Negative feedback is a message that stimulates an adjustment, the way that a rise in temperature causes a thermostat to turn on an air conditioner. A production process that begins to produce measurements that are outside its statistical control limits triggers an analysis and adjustments to the system. There is, however, another kind of information created in a system that may precipitate more than adjustments. This sort of information creates changes of an unanticipated or even cataclysmic dimension because it destroys or seriously impairs the self-regulating capacity of the system.

Reorganization is an increasingly recurring event in organizations. It may be that reorganization is proof that we are living in chaos or that the only certainty is uncertainty. My impression, however, is that most reorganization results from a lack of systems thinking on the part of leaders. In particular, reorganization reflects an indifference or unawareness to the connectedness of all the elements in an organization.

I observed a large-scale reorganization a few years ago in which a governmental agency gathered numerous small contractors into a single contract with the justification that it would lead to greater efficiency, higher quality, less cost, and the need for fewer government people to oversee the contract. What actually happened was

- Loss of direct influence of the government over the performance of the contractor.
- Breakdown of an effective, informal information and management network that had developed among the smaller contractors and their government counterparts.
- Creation of massive uncertainty about responsibilities and degrees of authority.
- Increase in costs.
- Loss of competence among government workers who were removed from hands-on job responsibilities to those of monitoring contractor performance.

The point of this illustration is that the organization is a system composed of very complex networks of communication that create predictable, unpredictable, useful, and destructive results. Grasping this concept of connectedness is fundamental to managing an organization as a system that exists in a systemic relationship with the earth's environmental system.

Information created by certain actions or events may create a new effect or compound an existing problem in very surprising ways. The creation of chloroflurocarbons (CFCs—trade name, Freon) was viewed as a major breakthrough for industry. CFCs initially were viewed as being of tremendous utility and were heralded as a significant environmental improvement because the common refrigerants at the time (ammonia, methyl chloride, and sulphur dioxide) were not suitable because of their noxious and toxic properties. CFCs became the compounds of choice for refrigeration, air conditioning, wind tunnels, and as cleaning agents in the communications and electronics industries. In the 1940s, they were introduced as aerosols. Their capacity to interact with ozone and to deplete this essential compound in the ionosphere was never anticipated. It is not that the impact of CFCs on the ozone layer above the earth could not have been anticipated. It was not anticipated because of a lack of systems thinking and analysis.

The DDT flow through the food chain described by Rachel Carson in *Silent Spring* is another example of connectedness and unanticipated results. The use of DDT caused changes in the food chain that impaired the capacity for self-adjustment and tended toward the extinction of certain life forms. Birds were not the targets of DDT, but they became the unanticipated victims.

A further example is the emission of CO_2. One possible scenario is that increased quantities of CO_2 caused by humans and their enterprises will lead to a rise in the earth's average temperature, which will lead to complex changes in the earth's ecosystems that are only partially understood. These changes will be too rapid for forests to regrow. The migratory patterns of birds will be altered with unknown consequences. Life in lakes that depend on certain temperature ranges will be destroyed. Climate change could, in short, intensify the pace of species extinction. It also could cause rises in sea levels, loss of wetlands, costal erosion, increased frequency and severity of storms, and probably many more problems that we cannot yet imagine.

Principle Two

> Sustainable performance is an ecologically interdependent process and requires that all organizational processes, products, and services be revised or replaced to ensure their compatibility with nature's ecosystems [10].

This principle follows logically from the first principle. It affirms the consequences of thinking of the organization as an open system that exchanges input and output with the environment. The requirement to revise every aspect of an organization's performance to reflect this exchange has been discussed in the previous chapter.

This principle affirms that a business system should function as an analogue to nature's ecosystems. The goal is to use the resources of nature's ecosystems in such a way that what is returned to these systems can be absorbed by and used to nurture those systems and not damage them. This cycle of benign use and return may involve many iterations of return and reuse within an organization's work processes [4].

Principle Three

> Sustainable performance is a results-oriented process and requires the demonstrated commitment of organizational leaders to specific, measurable results.

SP is a results-oriented initiative. Planning, involving the work force, developing partnerships, and every other activity associated with SP are intermediate events and means to the ends of improving environmental efficiency or performance and competitive position. It is essential, therefore, at the very beginning of an organization's awakening to the prospect of SP that it develop a very clear view of the results that it expects to achieve. The actual statement and communication of such a view is facilitated by developing a rationale and by developing a set of specific improvement goals.

Rationale. Much material for a rationale for SP already has been presented. The essential elements in the rationale are as follows:

1. SP is an inevitable response to the environmental challenge; it cannot be avoided.
2. SP offers the same opportunity for improved competitive position as did TQM and continuous improvement.
3. The sooner that organizations embrace SP, the more likely it is that they will maintain or improve their competitive positions.

This rationale must, of course, be translated to match the specific conditions of any given organization and the specific competitive needs of the organization. Some of the pressures that I have already identified will be more important to one organization than to another. Some businesses, such as food service, banking, consulting, training, and retail sales will not feel the same imperatives for change from environmental regulations as the chemical, manufacturing, utilities, and petroleum industries will. Also, some organizations will experience more pressure than others will from their greener competitors.

Organizations that already have highly developed social consciences and histories of responsible social behavior will have little trouble developing rationales for responsible environmental behavior. Whether the task is easy or hard, a rationale must be developed that removes any doubt that SP is a survival priority and that it must have the full commitment of every person in the organization—especially, of course, its key leaders and decision makers.

Improvement Goals. Improvement goals can be focused on any or all of the following opportunities:

- Environmental performance of suppliers and the environmental quality of their input.
- Environmental efficiency of work processes for using and transforming input.
- Environmental efficiency of output of services and products and their use by customers.
- Environmental performance of customers and the environmental efficiency of post-customer output.
- Output of waste, pollutants, and toxic emissions into the environment.

There also are at least two other areas of opportunity for improving environmental performance that are not discussed in this book in any detail. This book focuses on the primary processes of production. The other two opportunities for improving environmental performance are the work environment and nonwork-related activities.

The work environment includes employee safety, ventilation, noise pollution, and so on. Nonwork-related activities include eating, drinking, and the use of recreational facilities. A thorough commitment to SP will lead organizations to address opportunities in both these areas.

Here are a few of the work-environment pollutants that organizations may be called on to reduce or eliminate:

- *Formaldehyde*, which is emitted from plywood, carbonless copy paper, ceiling and floor tiles, paper towels, and perfumes.
- *Radon*, which seeps through cracks in the foundations of buildings.
- *Ozone*, which is emitted from copying machines and printers.
- *Asbestos*, used in pipe lagging and insulation.
- *Electromagnetic radiation*, which is emitted from electrical systems.

It is important that environmental-improvement goals be concrete and represent a significant change that will be achieved within a certain time period. Such goals will read somewhat like the following:

By x date:

- Reduce solid waste by x amount.
- Eliminate x pounds of packaging in x products.
- Reduce by x wattage the electrical energy used in x processes or facilities.
- Recycle x barrels of oil.
- Reduce by x percent the release of CO_2 from x.

- Replace all chemical cleaning with aqueous processes.
- Transfer x percent of petroleum-energy source to solar.

The third principle of SP requires organizations to focus on results. This involves at least two related actions: developing a rationale for action and developing a set of concrete improvement goals. Goals communicate to the work force what must be done, and the rationale communicates why these goals must be achieved.

Principle Four

Sustainable performance is a community-building process. This means that organizations need to cooperate with one another and use the environment in ways that are equitable for one another. It also requires that organizations involve all their stakeholders in the processes of planning and implementing sustainable performance.

The environmental challenge probably will have more practical effect on the development of a global community than any other event that has shaped the nature of human society and the relationships of nations. Even the threat of nuclear war was never sufficiently powerful to force people to believe that their own best interests were served by disarmament and the strengthening of international law or that their own best interests lay in becoming a global community. There always have been beliefs (such as "reason will prevail," or "sufficient force will keep war from happening," or "it is possible to defend against a nuclear attack") that have given people mental "ways out." In the U.S., people were encouraged to discount the threat of nuclear war by the promise of an electronic screen in space (a Disneyland-like fantasy).

Although the threat of nuclear war failed, the environmental threat must inevitably succeed. At the political level, international cooperation in response to the environmental challenge is growing at an extremely fast pace. This pace is mirrored in the world of business, as cooperation on environmental issues and technology is becoming more and more common. The growth of a communal response to the environmental challenge is founded on nothing more complicated than self-preservation.

It is clearly in the economic and survival interests of businesses to cooperate in response to the environmental challenge. The practice of such cooperation is evidenced in the many national and international coalitions that have developed over the past few years and which have been mentioned previously in this book. One excellent example of such cooperation is the chemical industry's Responsible Care Program.

The Responsible Care Program for corporate and environmental management was introduced by the Canadian Chemical Producers' Association and later adopted by the chemical industries in the U.S. and U.K. Participation in the Responsible Care Program is mandatory for organizations to hold membership in the Chemical Manufacturers' Association in the United States. Organizations that join the Responsible Care Program pledge themselves

- To recognize and respond to community concerns about chemicals and our operations;
- To develop and produce chemicals that can be manufactured, transported, used, and disposed of safely;
- To make health, safety, and environmental considerations a priority in our planning for all existing and new products and processes;

- To report promptly to officials, employees, customers, and the public information on chemical-related health or environmental hazards and to recommend protective measures;
- To counsel customers on the safe use, transportation, and disposal of chemical products;
- To operate our plants and facilities in a manner that protects the environment and the health and safety of our employees and the public;
- To extend knowledge by conducting or supporting research on the health, safety, and environmental effects of our products, processes, and waste materials;
- To work with others to resolve problems created by past handling and disposal of hazardous substances;
- To participate with government and others in creating responsible laws, regulations, and standards to safeguard the community, workplace, and environment;
- To promote the principles and practices of Responsible Care by sharing experiences and offering assistance to others who produce, handle, use, transport, or dispose of chemicals [11].

The Responsible Care Program not only acknowledges the need for cooperation among companies but it also recognizes the need to create cooperation with a company's stakeholders.

A stakeholder is any person or group that is affected by the performance of the organization, i.e., how it manages the system for sustainable performance and its many inputs and outputs (Figure 4-1). Identifying stakeholders for small, single-site organizations will not be difficult. For large organizations with multiple sites and multiple operations, the task can be a formidable one.

Stakeholders, for purposes of planning SP policy, goals, and strategies, may extend far beyond the traditional list of employees, shareholders, lenders, lawmakers, enforcers, suppliers, and customers. Organizations are finding more and more that "birds don't vote, but bird watchers do." Today, stakeholders include environmental groups, citizen activists, coalitions of trade and professional groups, contractors, waste-management companies, and a variety of others.

An organization should try to include all of its stakeholders on its environmental team. Stakeholders can be important sources of information about current and future issues. They can be valuable contributors to the solution of problems, and they can greatly reduce the number of future problems and issues that the organization may face.

Involving stakeholders in the design and implementation of an organization's SP initiative may not sit well with more traditional executives and managers. There is, however, no choice. The pressures for change that must be managed in the environmental age are all represented by stakeholders. Neither these pressures nor the stakeholders can be ignored. An early decision in implementing SP is to determine who the organization's stakeholders are and how to involve them in the SP initiative. The stakeholders of organizations will be involved—either positively and productively because they have been invited to be members of the environmental team or negatively and destructively by default.

As an organization begins to develop an awareness of its impact on the natural environment and begins to understand the kinds of pressures that it must manage, it will recognize that any movement toward SP invariably must be a communal one. It is impossible to undertake either SD or SP without recognizing that the environment binds all humans and all human business enterprises together. The principle of community will lead organizations to build coalitions and bridges with other organizations, governments, and academic institutions. This principle also will lead them to develop higher levels of communication, cooperation, and involvement with all their stakeholders.

Principle Five

> Sustainable performance is a limiting process. It requires that organizations recognize that there are costs associated with the earth's resources and ecosystems that must be included in the organizations' accounting processes and that will place limits on the size and nature of their businesses.

This principle of sustainable performance emphasizes the idea that organizations must manage and improve performance within tolerances or parameters established by the environment. Specifically, this principle requires that organizations recognize the limits of the earth's resources and the limits of resiliency in the earth's ecosystems. SP affirms that organizations must recognize that there are costs associated with the earth's resources and ecosystems. These costs must be included in their accounting processes and that will place limits on the size and nature of their businesses.

The environment does not place limits on profit. Indeed, an aggressive response to the environmental challenge carries many more payoffs than liabilities, and the payoffs outweigh the costs. In addition to the more obvious payoffs such as cost savings, cost avoidance, and risk reduction, by embracing SP an organization can expect to achieve some other (not so obvious) payoffs such as

- More open, creative, and responsive management because SP can be managed only by surfacing problems and aggressively pursuing new solutions.
- Improved labor relations, with improved health and safety standards that are inherent in improved environmental performance.
- Greater control of its own destiny by taking the offensive and going beyond compliance to SP.

There is a limit to the earth's resources. There are obvious limits to the amount of water that can be made available for agricultural and industrial uses. The availability of nonrenewable resources such as coal and petroleum is finite. The increasing tendency to switch to natural gas will make unprecedented demands on this source of energy. The exhaustion of mineral resources does not seem at present to be near, although the way in which minerals are mined, refined, and used does present major threats to the environment because of the wastes and pollutants that are produced.

It is precisely the issues of pollution and wastes that define the ways in which organizations must develop and which set limits to their development. SP is limited by the stress that waste and pollution place on the earth's ecosystems. Three critical areas of concern for industry are the emissions of carbon dioxide, sulphur dioxide, and nitrogen oxides.

Carbon dioxide (CO_2) is the primary contributor to the greenhouse effect and global warming. CO_2 emissions have increased by 10 percent over the last twenty years and are predicted to increase even more radically in the next ten years. Approximately two-thirds of the CO_2 released into the atmosphere can be attributed to human activities, particularly fossil-fuel combustion. One third of fossil-fuel combustion comes from the generation of electricity. Thus, the emission of CO_2 places a limit on SP.

The concentration of chlorofluorocarbons (CFCs) has increased dramatically in the last fifty years. CFCs are the primary cause of the "hole" in the earth's ozone layer and the resulting threat of damage to life from ultraviolet radiation. Higher levels of ultraviolet radiation contribute to increased frequencies of cataracts and skin cancer and may depress human immune systems. Higher levels of ultraviolet radiation also reduce crop yields, deplete marine life, and increase smog. The continued use of CFCs and the release of chlorine atoms into the atmosphere places another limit on the development of SP.

Two more emissions that place limits on the way business must function in the environmental age are sulphur dioxide (SO_2) and nitrogen oxides (NO_x). The quantity of these compounds released into the atmosphere has greatly increased in recent years. These pollutants are the main causes of increased acidity in lakes, rivers, forests, and soils. They also contribute to the deterioration of stone and metal structures. These gases are produced mainly by the combustion of fossil fuels. Their emission places still another limit on SP.

A final example of a limit relative to the environment is toxic waste. Toxic chemicals and heavy metals from industrial and agricultural use build up in soils and sediments and find their way into rivers, lakes, and oceans. These wastes are a direct threat to marine life. They present a much wider threat as they enter the food chain and become more concentrated and toxic as they pass up the chain, or as one organism or animal becomes food for another.

Continued reduction of the availability of energy and material resources and the lessening capacity of the environment to accept pollutants, toxic emissions, and waste place limits on how organizations can function. It is a principle of sustainable performance that organizations must learn to function within these limits.

Principle Six

> Sustainable performance is an open process and requires that organizations communicate fully all aspects of their planned and actual environmental performance to all organizational stakeholders [1].

Two-way, timely, and concrete communication with all stakeholders is essential for organizations to develop the information required for the continuous improvement of SP. However, acting from this SP principle does not mean acting to achieve an image-building public-relations goal or to gain positive visibility with internal and external audiences.

When customers, suppliers, the immediate community, and the many other stakeholders of an organization are treated as *audiences*, the communication process is reduced to the task of pleasing and not of giving and receiving useful information. The Community-Right-to-Know Act and various other statutes, regulations, and agreements remove any possibility that organizations can successfully manipulate environmental information. If any person or group persists, they will get the information they want about the environmental performance of an organization. Much of the relevant information, at least for the S&P 500 companies, is compiled regularly and updated by the Investor Responsibility Research Center.

The purpose of communication is improvement in the environmental performance of an organization. The following are the characteristics of communication that contribute to continuous improvement:

1. Purposeful
2. Inclusive
3. Two-way
4. Concrete
5. Timely

Purposeful. Communication is for the purpose of improvement. It must stay focused on two activities: (1) retrieving information that can lead to new and better ways of managing the whole system of SP, from presupplier input to postcustomer waste; and (2) providing information to stakeholders to build them into the organization's team.

Inclusive. Information loops and networks must be established across all internal and external interfaces. This requires not only identifying all the organization's stakeholders so that no person or group is left out of the information process, but also requires the use of a wide variety of methods (formal and informal) for communicating across these interfaces.

Two-Way. Dialogue is the process necessary to identify and clarify needs, to identify and resolve problems, and to support learning and innovation. Two-way communication can be assisted by actions such as including environmentalists on organizational boards and committees; establishing organizational and community environmental-improvement teams; and making provision for receiving and processing environmental questions through hotlines.

Concrete. Communication about environmental performance, like communication about any other kind of performance, must be in terms of specific goals and measured achievement. The third principle of SP requires that an organization establish specific improvement goals. As time goes on, performance against these goals can be tracked and measured. The fourth and sixth Principles of Sustainable Performance require that such performance information be widely disseminated through annual reports, special environmental reports, stakeholder briefings, and the like.

Timely. Timeliness as a quality of open and complete information exchange underscores the ideas of utility and no surprises. The goal is for an organization to be so responsive to its stakeholders that it anticipates its stakeholders' needs and concerns. By achieving this goal, organizations will develop stakeholders who will help it to anticipate environmental issues and find ways to resolve those issues that are mutually acceptable.

The Proctor & Gamble Company makes a concerted effort to communicate its environmental policies, goals, and initiatives to its stakeholders. It uses newsletters, its annual report, and a variety of other special publications—including an excellent leaflet, entitled *Total Quality Management: A Systematic Approach to Continuous Environmental Improvement*. Particularly noteworthy are the publications produced by P&G's Environmental Safety Department (ESD). The ESD's mission is to develop and test methods to assess the environmental safety of the company's products and their ingredients.

Principle Seven

Sustainable performance is a process of continuous improvement of every aspect of an organization's performance; it requires the full involvement of every member of the work force [7,8].

The continuous improvement of total quality is largely a function of employee involvement through team formation and development [7,8]. SP is the next evolution in total quality management. It is not surprising, therefore, that organizations that are reporting successes in environmental performance frequently acknowledge their dependence on teams and employee involvement for these successes.

Central to Anheuser-Busch's environmental initiative, for example, has been the establishment of committees that represent a cross section of employees and management. A group of Anheuser-Busch employees made a thorough analysis of solid-waste materials that were sent from the brewery to landfills. They identified materials that were recyclable and set up programs to recover these materials before the waste was sent to the dumpsters. This led to a 22 percent reduction in solid waste sent to landfills from the Baldwinsville, New York, brewery. [10]

The continuous improvement of quality is a function, to a very large degree, of the mental resources that are focused on improvement. The waste of mental resources is the primary and most critical resource issue that organizations face. Organizations do not run on technology or electricity or information; they run on the minds and mental energy of the people in them.

SP requires that organizations develop specific strategies and support systems to involve people in the processes of inquiry, problem solving, and innovation.

The terms involve and involvement are not particularly descriptive. The goal of involvement is to make maximum use of the mental resources of the entire work force. "Empowerment" is a term that has gained considerable popularity because of the need to give more content to the idea of involvement. Empowerment carries such meanings as follows:

- "Pushing authority for decisions as far down the organizations as possible."
- "Letting the people closest to a problem solve the problem."
- "Giving people a job and staying out of their way so they can do it."
- "Increasing the sense of ownership that people have for their work and their organization."

The most consistently successful strategy for building empowerment into a work force is team development [7,8]. The primary opportunities for improving SP are evident from the Systems Model for Sustainable Performance (Figure 4-1). Teams, in the form of natural work groups, process-improvement teams, quality-improvement teams, and so on, provide the best opportunity to develop and use the mental competencies of people in improving the total environmental performance of the organization.

Principle Eight

Sustainable performance is a data-based process and requires concrete information retrieved from auditing, measuring, and reporting the organization's environmental performance [1].

This principle is related to Principle Three and emphasizes the technical aspects of building data bases by which the continuous improvement of SP can be planned and undertaken.

The International Chamber of Commerce has defined environmental auditing as a

...management tool comprising a systematic, documented, periodic and objective evaluation of how well environmental organization, management and equipment are performing with the aim of helping to safeguard the environment... [5].

SP stretches across every dimension of an organization's life. SP, like all quality initiatives, must be managed by data. To be useful, data must be reliable and must be communicated. Auditing and reporting ensure the possibility of managing SP by ensuring that reliable data are gathered and communicated. The concept of auditing, in practice, takes various shapes and serves a number of purposes. These will be discussed in detail in Chapter 7.

Principle Eight carries the implicit requirement that auditing and reporting be sufficiently pervasive and periodic to achieve the following purposes:

- Demonstrate the organization's practical and performance-based commitment to SP.
- Provide a broad base of environmental information for planning and improving SP.
- Verify compliance and noncompliance.
- Reduce risk and exposure to litigation.
- Increase employee involvement in the continuous improvement of SP.
- Assess training programs and training needs.
- Support rewards and celebrations for improvements in SP.
- Assess the relative performance of different organizational sites and elements.
- Identify opportunities for cost avoidance, cost savings, and improved profit.

A reporting system can be designed around various outlines or headings, including the principles outlined in this chapter. Such a system should cover at least the following topics:

- Restatement of SP policy and goals
- Action to reduce solid waste
- Action to reduce hazardous waste
- Action to reduce pollution
- Reduction of energy use and employment of alternative sources of energy
- Environmental improvement of services and products
- Management of risk
- Relationship with stakeholders

Principle Nine

Sustainable performance is a technologically dependent process and requires organizations to develop partnerships with governments, other businesses, educational entities, research and development groups, suppliers, and customers in order to discover and implement ways to improve sustainable performance [2,3].

SP is the practical outcome of the growing recognition that the environmental challenge is global and critical. It is not possible for any one business or group of businesses to manage the stakes and the pressures of the challenge, any more than it is for one nation or a group of nations (developed or developing) to manage the challenge. The nature of the environmental challenge requires cooperation and information transfer on an unprecedented scale. Principle Four emphasizes the community-building characteristic of the environmental challenge. The technological challenges are too big for any one company or industry to solve. Because SP is heavily dependent on the development of new technologies and more creative ways to manage the problems of waste and pollution, it is imperative that organizations develop collaborative initiatives to create the technologies that all require.

Examples of cooperative industry groups are the Partnership for Plastics Progress, the Institute of Food Technologists, the Computer and Business Equipment Manufacturers Association, and the Institute of Packaging Professionals. Groups that are international in scope and involve the entire business community include the Global Environmental Management Initiative (GEMI), the Management Institute for Environment and Business (MEB), the Business Council for Sustainable Development (BCSD), and the Coalition for Environmentally Responsible Economies (CERES).

Business partnerships to develop new environmental technologies and their transfer will not ensure competitive position, but they will ensure that an organization can stay in the game. The reduction and elimination of CFCs, for example, affects a wide range of industries. The formation of the Industry Cooperative for Ozone Layer Protection (ICOLP) is a response to the recognition that the CFC problem must be resolved and that the sooner industry takes the initiative in resolving it, the more likely it is that it will escape more punitive and costly legislation. ICOLP's mandate is to sponsor and coordinate the open and worldwide exchange of information on ozone-depleting solvents in the electronics industry. One very impressive goal of ICOLP is to eliminate CFC solvents from Mexico's electronics industry by the year 2000.

The foundation of all substantive change is learning. In trying to develop a systematic and unified response to the environmental challenge, organizations become stymied by (1) needing to know more about what is already known; or (2) needing to investigate that which is not known.

Companies that commit to SP must commit to the process of continuous learning, the development of new technology, and the acquisition and use of new technology. The specific strategies for undertaking such a commitment include the following:

- Developing the potential for inquiry and learning
- Investing in research and development
- Forming networks and coalitions to transfer technology
- Training

SP is a technologically dependent process. The technology needed can be best developed through a variety of business, government, and research coalitions. SP also is a total process, and it requires the reformation of all the traditional management and human resource systems in an organization.

Principle Ten

> Sustainable performance is a total organizational process and requires that all planning, decision making, and human resource systems be made fully congruent with the organization's commitment to SP.

Environmental performance must be fully integrated into the organization's management practices and systems. It must be imbedded in strategic planning, in information systems, and in financial-management systems. People at every level must learn to perceive concern for the environment as "the right thing to do." Environmental performance must be expected, acknowledged, rewarded, and celebrated.

One of the general weaknesses in the way TQM has been conceived and initiated has been the failure to revise and restructure organizational planning, business, and personnel systems to support TQM. For example, the underlying strategy in TQM to build the potential for continuous improvement is team formation and team development [7,8]. One persistent failure, however, is that organizations do not take into account the need to build systems to support team development and performance.

It is obvious that to develop teams, we need to get people (in the many different ways that they may be related in an organization) to work as teams. This means, first, that we build teams wherever people are already naturally related in organizations, e.g., as work groups, as managers and staffs, as project groups, or as committees. Developing teams also involves organizing new teams such as quality-action teams, process-improvement teams, interface teams, and network teams.

Viewed from an organizational perspective, however, team development is more than developing teams from already existing groups or organizing new groups of people into teams. If teams and teamwork are to become dominant and lasting characteristics, organizations must use team-development strategies that go beyond organizing teams and training people to function as teams. The kinds of strategies that are required to develop the whole organization as a team include the following:

1. Management-team modeling
2. Strategic planning for team development
3. Team-focused organizational surveys
4. Team-centered training
5. Team-centered selection
6. Team-centered rewards and appreciation

It is apparent that the same sort of strategies used to revamp organizational systems to support team development must be used to support SP. In organizations that commit to SP, managers must "walk the talk" of environmental performance. Environmental priorities must be highlighted in the strategic plan, and every aspect of the organization's selection, reward, and promotion practices must reflect its commitment to SP.

Summary

Sustainable performance represents a radical shift in the values of an organization. These values can be summarized by ten principles. The Principles of Sustainable Performance complete the description of SP in this book. In the next chapter, I will begin to discuss the decisions and actions that leaders must take in responding to the environmental challenge and in managing their organization's journey toward SP.

The Sustainable-Performance Management Model (Figure 1-1) shows the process of managing sustainable performance along a path that leads from formulating an environmental policy and ends with improved performance. The feedback loop indicates that this process continues until SP has become the normal way of doing business and no longer has to be considered a special initiative.

Managing the path that leads through the various mile stones of the model requires input from various information sources. Thus far, I have discussed the pressures and the sources of information related to the meaning of SP.

Organizations will respond to the environmental challenge in ways that reflect the sensitivity of management to the issues, the kind of motivation that is characteristic of the organization, and the competencies of the people in the organization. In the following chapter, I will describe these response levels.

References

1. *Business Strategy for Sustainable Development.* (1992). Winnipeg, Canada: International Institute for Sustainable Development.
2. Elkington, J., & Knight, P., with Hailes, J. (1991). *The Green Business Guide.* London: Victor Gollancz, Ltd.
3. Elkington, J, & Burke, T. (1987). *The Green Capitalist.* London: Victor Gollancz, Ltd.
4. Frosch, R., & Gallopoulos, E. (September, 1989). Strategies for Manufacturing. *Scientific American,* pp. 94-102.
5. International Chamber of Commerce. (1989). *Environmental Auditing.* Paris: Author.
6. Kinlaw, D. (1989). *Coaching for Commitment: Managerial Strategies for Obtaining Superior Performance.* San Diego, CA: Pfeiffer & Company.
7. Kinlaw, D. (1992). *Continuous Improvement and Measurement for Total Quality: A Team-Based Approach.* San Diego, CA: Pfeiffer & Company.
8. Kinlaw, D. (1991). *Developing Superior Work Teams: Building Quality and the Competitive Edge.* San Diego, CA: Pfeiffer & Company.
9. Kinlaw, D. (1992). *Team-Managed Facilitation: Critical Skills for Developing Self-Sufficient Teams.* San Diego, CA: Pfeiffer & Company.
10. Long, L. (1990). Measuring Anheuser-Busch's Environmental Goals. *Conference Proceedings, Corporate*

Quality, Environmental Management II. Washington, D.C.: Global Environmental Management Initiative, pp. 83-86.

11. Responsible Care Program Brightens Chemicals Image. (October 1, 1990). *Chemical Marketing Reporter*, p. 7.

12. World Commission on Environment and Development. (1987). *Our Common Future.* New York: Oxford University Press.

6

RESPONSE LEVELS

The Sustainable Performance Management Model (Figure 1-1) indicates that SP is a process that requires a significant amount of new information in order to be managed successfully. The sources of this information are

- An understanding of the pressures working on the organization.
- An understanding of the characteristics of sustainable performance.
- Use of the Systems Model for Sustainable Performance.
- Use of the Principles of Sustainable Performance.

An additional source of information that can assist in managing the milestones to SP is the *level of response* that the organization is making and able to make to the environmental challenge.

SP, like every other major change in an organization, requires a host of management decisions and actions to build the capacity of the organization to function in a new way. These decisions and actions must be based on a clear understanding of the pressures forcing organizations toward sustainable performance and a clear understanding of SP. In addition, leaders must:

- Understand the levels of response that organizations can be expected to make to the environmental challenge.
- Know what level of response their own organization is making.
- Be able to influence the level of response their organization is making and lead it toward SP.
- Know the milestones their organization must reach in its progress toward SP.
- Understand and use the general strategies for SP.
- Understand and use the special tools and technologies of SP.

In this chapter, I will discuss managing the level of an organization's response to the environmental challenge.

The Response Model for Sustainable Performance

Organizations are responding to the environmental challenge in different ways. The responsibility of leaders is to understand how their organizations are responding and to develop plans and strategies for moving their organizations toward the highest level of response: SP. The Response Model for Sustainable Performance (Figure 6-1) is a tool that leaders can use in thinking through their responsibilities for managing the environmental performance of their organizations.

Response Levels / 193

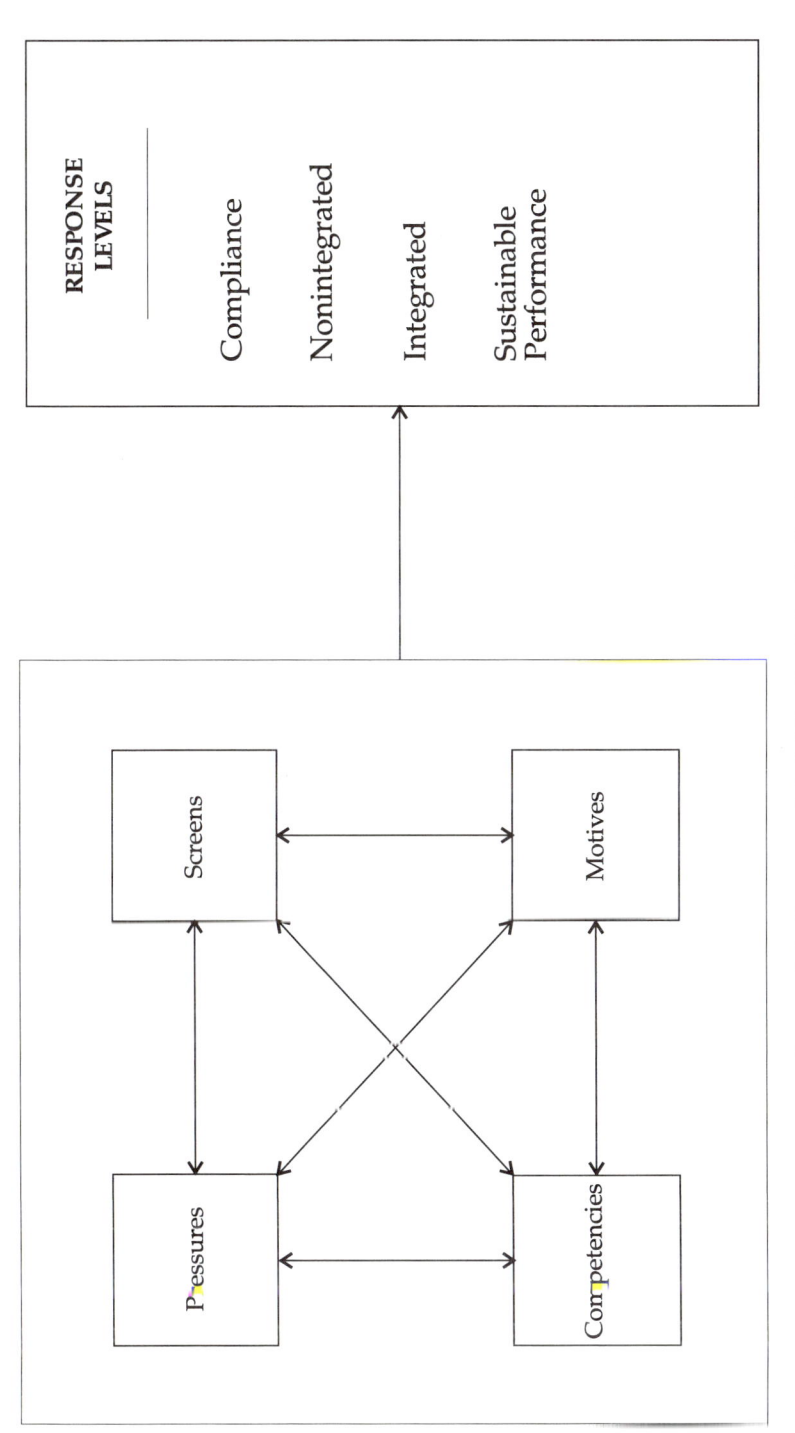

Figure 6-1. Response Model for Sustainable Performance

Elements

The elements of the model are pressures, screens, motives, competencies, and response levels.

1. Inexorable **pressures** are working on organizations that are forcing them to respond to the environmental challenge and requiring them to change the ways in which they do business.
2. The way that leaders respond to these pressures—and to the severe threat to the environment that underlies them—will be significantly influenced by the **screens** that leaders use to the filter information that they receive about the pressures and the environment.
3. The degree to which and way in which an organization responds to the pressures and the ease with which it moves toward SP will be influenced by the different kinds of **motives** that are operating in the organization.
4. The **competencies** that the people in the organization have for responding to the environmental challenge and working toward SP also will affect the organization's level of response.
5. The interactions among the pressures, screens, motives, and competencies will determine the actual **level of response** that an organization is capable of making to the environmental challenge.

Leadership Tasks

The Response Model for Sustainable Performance suggests certain tasks that leaders must perform and questions that leaders should ask as they begin to think about moving their organizations toward SP. These tasks and questions are as follows:

1. Understand the pressures that require organizations to change to SP as their overarching business philosophy. (Do I understand the pressures that are working on my organizations and all organizations in this environmental age?)
2. Understand and manage the screens that they use to filter the information about the environment that they receive and use. (Do I understand my own ability to receive and understand information about the environment? What may I be doing to eliminate or distort the information I receive about the environment?)
3. Understand SP and help the organization to integrate the principles and characteristics of SP into its policies, goals, strategies, and practices. (Do I understand the framework within which improved environmental performance should be planned and the level of change that is required to make such improvement?)
4. Understand and manage the motives that will energize their organization's response to the environmental challenge. (Do I understand the motivations of myself and the rest of the people in my organization for responding to the pressures and the challenge?)
5. Ensure that the work force has the necessary knowledge and skills to be fully involved in SP. (Do I know the competencies that I and the rest of the people in my organization need to undertake the process of SP?)
6. Understand their organization's current level of response to the environmental challenge in order to lead the organization toward SP. (Do I know

enough about my organization's current level of response to the environmental challenge to develop a plan to improve that response?)

Pressures

The cornerstones of the Response Model for Sustainable Performance are (1) the *sources* of the response (the *pressures*); and (2) the *goal* of the response (*SP*).

The pressures on organizations are described in Chapter Three. These pressures are indifferent to the will of individual organizations, and the survival of any one organization will depend on its capacity to respond to these pressures in time to assure its own competitive position.

The response that an organization makes to the pressures for change and the stakes that underlie these pressures are influenced by at least three elements: the mental screens that decision makers use to filter information about the environment; the motives that are behind the organization's response to the environmental challenge; and the competencies that people in the organization have for being actively involved in SP.

Screens

The first variables that determine the kind of response that an organization will make are the screens that organizational leaders employ in receiving and using information. It is important for leaders to understand these screens and to reduce the influence that they have over the leaders' perceptions and actions.

The screens that appear to have the most influence on the kinds of responses that leaders make are

- The general awareness and sensitivity that leaders have to the severity of the environmental challenge.
- The degree to which leaders perceive environmental action as serving the organization's best interest.
- The time horizons that leaders use.

Screens are conditions that exist in the minds of organizational decision makers. These screens influence how leaders respond to the pressures acting on the organization and strengthen or weaken their motivation to take positive action.

Management Awareness

Awareness results from (1) the amount of information that leaders have about the environment and environmental pressures; and (2) the level of competence that they have to evaluate and use this information. What leaders know about and what they believe to be important determine where they focus their attention and how they use their time.

For example, leaders in the U.S. traditionally have believed that the future of their organizations depends on immediate profit and fast turnaround on investment. They have, therefore, acted as executive accountants and have tried to manage their organizations through budgets, reports, and figures rather than through paying attention to their customers and the quality of their services and products.

This focus on profit has led managers to adopt a formula for profit that looks like this [4]:

SELLING PRICE = COST TO DELIVER QUALITY + PROFIT

The tradition is to charge the customer a price that is computed *after* the amount of profit desired is added to the cost to deliver the service or product. The selling price to the customer is the last consideration. Desired profit margin and cost to deliver are computed first. This formula may work well in environments that are free from competition. It is a formula for disaster in an age in which the most competitive organizations think first of the customer. The traditional formula does not encourage managers to worry most about quality and, at the same time, to figure out how to reduce the cost to deliver a service or product.

The profit formula that puts the customer first and stimulates continuous improvement is

SELLING PRICE − PROFIT = COST TO DELIVER QUALITY

The focus in this formula is, first, on what the market price for a quality product or service can be and, second, on achieving profit by relentlessly reducing the costs to deliver that service or product. The primary way to make a profit in a highly competitive market is to reduce the cost to deliver. It is not to raise the selling price.

Another example of misplaced emphasis is apparent in the problem-solving behavior of leaders. Too often they pay attention to solving problems that have occurred rather than to improving what is problem-free.

Perhaps the most important example of misplaced time and attention is in the degree to which managers and other organizational leaders avoid spending time in learning. A brief list of what many organizational leaders have not learned includes the following:

- That superior performance can be achieved only through the commitment of people, not by control.
- That the one critical resource in any organization is the mental capability of its people.
- That teams are the fundamental unit of performance, and that managing teams, not managing individuals, is the key to superior performance.

- That they are not paid to make decisions but are paid to make the best decisions.
- That creating a climate for continuous learning is the key to continuous improvement.
- That, above all else, they must be model learners.

As I already have noted, there is a realization among a number of senior executives and managers of the environmental challenge and the imperative to move quickly toward a new way of doing business. However, this sense of urgency is not pervasive among senior executives, and most managers are still "doing business as usual."

As long as managers at all levels are not fully aware of the environmental stakes and pressures that are shaping the future of business and as long as they do not understand the competitive challenge contained in the concept of SP, they will not aggressively lead their organizations to establish clear environmental policies, set significant improvement goals, and undertake major programs to reduce, recycle, and reuse.

Self-interest

A second screen that affects the way in which managers view environmental stakes and pressures and their motivation to undertake the improvement of their environmental performance is their own self-interest. This screen establishes how closely managers see their environmental performance aligned with their organization's bottom line and their own personal success.

Politicians provide good examples of how the perception of self-interest influences what people say and do. Most do not perceive their own political success (i.e., self-interest) tied to strong stances on the environment. Their typical stances are something like: "I will not be remembered as the person who sacrificed American jobs for an environmental issue like the Spotted Owl, the ozone hole, spiders in the rain forests, CO_2 emissions...."

Senior U.S. managers historically have not been great risk takers. For years, automobile-manufacturing executives in Detroit lived by the principle, "never be the first one to try something new." Their self-interests were not served by being innovators. As one manager I interviewed put it, "Ours is a closely held company. The owners will raise merry hell if I start proposing that we cut into their profits by going beyond what the EPA wants."

Environmental disasters in organizations tend to increase the alignment between environmental performance and self-interests. Disasters such as the toxic-chemicals release in Bhopal, the Rhine River oil spill, and the Exxon Valdez oil spill caused many organizations to review their own environmental policies and practices and to change their internal staffing and environmental organization. In particular, these disasters emphasized the vulnerability of chemical and petrochemical companies.

Organizations that see their financial success tied to their environmental performance do not make distinctions between expenditures to reduce the time and materials associated with a process and expenditures to improve their environmental performance. The costs for energy, the costs to manage waste, the costs to monitor toxic emissions, and the costs to clean up contaminated industrial sites all are regarded as overhead costs. Any organization that understands that its own self-interest is served by reducing overhead will find no trouble accepting the emerging environmental technologies for reducing, reusing, and recycling as being aligned with its own self-interest.

Time Horizon

One of the consistently identified problems in improving the performance of organizations is that decision makers operate with a short time horizon. The emphasis in the United States consistently has been on quick turnaround of investment, immediate improvement in sales, and annual increase in profit. The reporting cycles that require quarterly and annual accounting suggest a good deal about the typical time orientation of managers.

The way that managers interpret information about the environment and respond to this information is determined, in part, by their time horizon. Elliot Jaques, director of the Institute of Organizational and Social Studies at England's Brunel University, estimates that most people cannot systematically think and plan beyond a three-month time horizon, and that only one person out of several million in the West can mentally encompass the next twenty years [3].

It is now common knowledge that the longer time horizon of the Japanese has been a critical variable in their successful domination of the consumer-electronics industry and the automobile industry. Although U.S. firms were the first to introduce video-recording technology, they refused to finance the long-term research and development costs that transformed this technology from industrial and professional use to general consumer use. The Japanese worked with a much longer time horizon and, as a consequence, they are now the unchallenged leaders in the most profitable part of the consumer-electronics marketplace.

The cost of capital probably has been a major deterrent in the ability of U.S. business leaders to develop a long-term time horizon. This cost often motivates business leaders to opt for a quick turnaround and quicker return on investment. Also, in the U.S., companies are much more likely to raise capital in the open market from the sale of securities than they are from financial institutions such as banks and insurance companies. The old saying "when banks lend you a hundred dollars it's your problem, but when they lend you a million dollars it's their problem" applies. When banks loan large amounts to corporations, they become actively interested in the management and performance of these organizations.

In Japan, when banks make large, long-term loans to organizations, they typically own equity in the organizations and participate actively as members of their boards. This arrangement and similar ones create a much greater possibility for long-term planning and commitment.

However, external circumstances have not created the U.S. preference for quick fixes and short-term gains. Managers have chosen their time horizons.

The ranks of senior executives and managers are still dominated by individuals with highly developed financial and legal skills. These individuals become easily divorced from their firm's products, services, markets, and means of production. They rely more and more on accounting information for making decisions and less and less on the competencies of people, the quality of their products, and the expectations of their customers. The rash of mergers and takeovers in the 1980s and early 1990s is sufficient testimony to this financial mind set.

Time horizon is a critical variable in determining how organizations will respond to the environmental challenge. As with investments in total quality management, there may be a considerable time lag from the point at which investment in the continuous improvement of quality is begun and the time when payoffs and return on investment are realized.

Environmental problems are systemic and cultural. The expectation of quick payoffs and immediate returns will lead organizations to focus on public relations and green marketing rather than on the long-term changes that will produce long-term continuous improvement and a firm position in the marketplace. The "end-of-pipe" mentality that has dominated environmental improvements and much of environmental technology is fueled by a short-term time horizon.

The screens that dominate the way in which organizational leaders process and use information about the environment greatly influence the kinds of responses they lead their organizations to make. These screens are awareness, self-interest, and time horizon.

Motives

Motive is the next variable in the Response Model for Sustainable Performance that influences the level of response that an organization will make to the environmental challenge. At least one other study about organizational response has tried to integrate actual responses with underlying motives [1]. This study defined motives as the causes that generated the environmental responses of companies (similar to my list of pressures). The "motives" identified were as follows:

- Economics
- Regulation
- Environmental disasters and accidents
- Corporate culture
- Reputation
- Leadership

I do not define motives as the external goals or conditions that explain action, as this study suggests. I define motives as the internal cognitive conditions that lead individuals and organizations to chose to perform in certain ways and not in other ways. In my use of "motives," I am proposing that organizations are all faced with the same pressures driving them toward SP. Part of the difference in the way they respond is the difference in what motivates one organization and the people in it to respond from what motivates another organization and the people in it.

Motives are labels that we give to certain performance phenomena in order to categorize these phenomena, to understand them, and to develop strategies for affecting them in some way. Motives are interpretations of what we observe and are ways of accounting for what we observe. Motives are answers to the question, "Why did the person do that?" or "Why did the organization do that?"

Organizations respond differently to the pressures driving them toward SP. By understanding the motives that lead to these responses, leaders can help their organizations to develop new motives or strengthen old ones.

We can identify at least three motives that appear to account for the ways in which organizations respond to the environmental challenge. These motives are *match, return,* and *expectation*:

- **Match** describes the degree to which individuals and organizations perceive that there is congruence between what they value and the goals or opportunities presented to them. Organizations that already perceive themselves as socially responsible will quickly accept the challenge to improve their environmental performance as appropriate and fully congruent with their already existing organizational image. Match is the measurement of *intrinsic* reward.

- **Return** describes what individuals and organizations expect to gain as material rewards or what pain they expect to avoid by undertaking some actions. Organizations that want to avoid punitive fines, litigation, and criminal liability will work to comply with environmental regulations. They will not necessarily work to go beyond the requirements of these regulations, unless they see that going beyond will lead to a better competitive position and more profits.
- **Expectation** refers to the motive to undertake what is perceived as achievable. This motive interacts with the first two to determine just how organizations will respond to the pressures driving them toward SP. Organizations may fully endorse the highest goals of SP, but their strategies will reflect what they perceive they actually can achieve within the limits of their resources and the constraints of their business objectives.

These motives are integral to the motivational model outlined in Figure 6-2 [5]. The model proposes that observed behavior or action results from

1. The *self's* evaluating goals and values and determining the degree of *match* that exists between perceived goals and the values of the self.
2. The *self's* evaluating the rewards that can be obtained and the costs that can be avoided and calculating the *return* that is expected.
3. The *self's* evaluating the resources and blocks and determining the strength of *expectation* associated with the probability of success.

206 / Competitive & Green

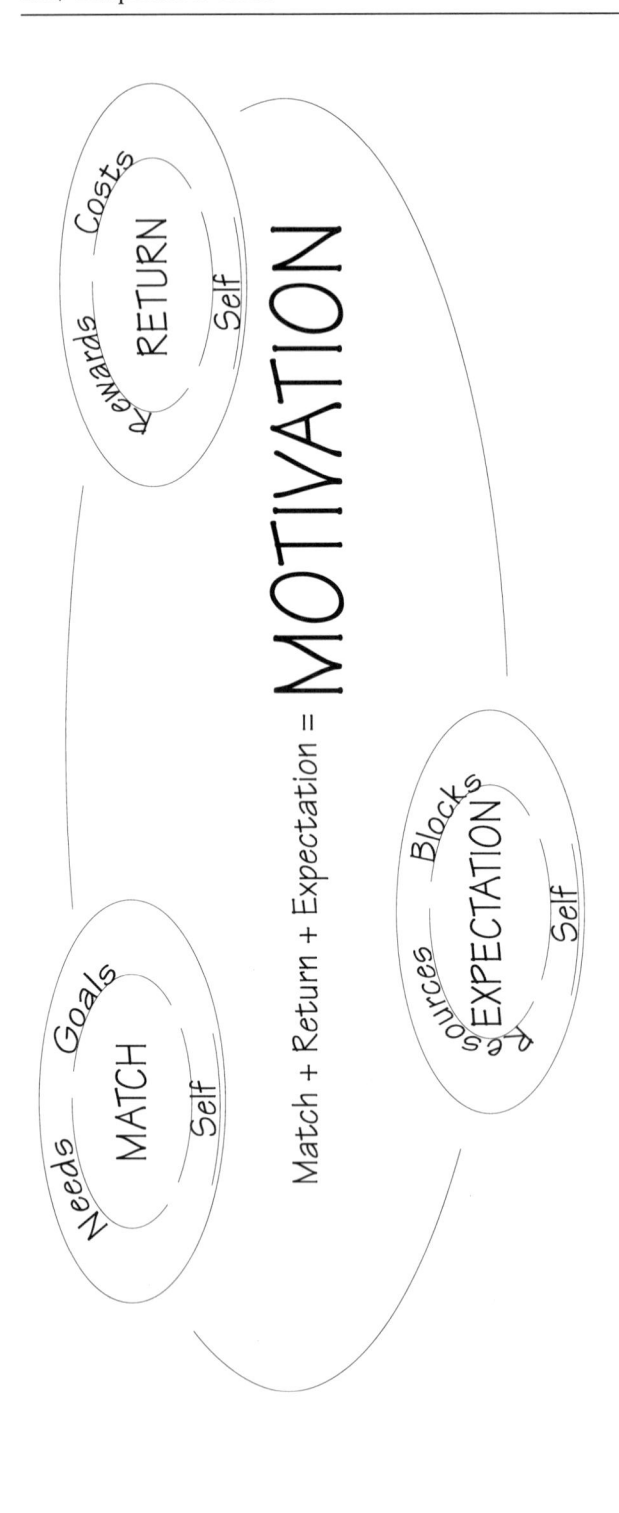

Figure 6-2. Motivational Model

Match + Return + Expectation = MOTIVATION

MATCH: Needs, Goals, Self
RETURN: Rewards, Costs, Self
EXPECTATION: Blocks, Resources, Self

Source: D. Kinlaw (1990). *Motivation Assessment Inventory*. Norfolk, VA: Kinlaw Associates.

Individuals who pride themselves on being law-abiding will stop at traffic signals, obey road-speed laws, and not cheat on their income taxes—regardless of whether they might be caught or punished for not obeying the law. Organizations that truly view themselves as honest will not cheat on contracts or overcharge customers—even if they know they will not be found out.

In the case of an organization that responds quickly to the environmental challenge, there is a dominant perception that this is what the organization ought to do. Considerations of profit are not placed before moral and ethical considerations, so concerns for profit are not placed before concerns for the environment. There is a *match* between the self-perception of the organization and the goals of responsible environmental behavior.

One very popular example of an organization that demonstrates a close match between organizational values and environmental performance is Body Shops International, directed by its founder, environmentalist Anita Roddick. In the time since she opened the first Body Shop in Brighton, England, Ms. Roddick has led campaigns to save the rain forests and the whales, to prevent testing on animals, and to stop abuses of human rights. Body Shops sell natural toiletries that are produced by indigenous people in developing countries. Nothing sold has been tested on animals. Stores are centers for environmental education, and all employees are given time to participate in environmental programs. Roddick and Body Shops International are dramatic proof that environmental responsibility and profit are not mutually exclusive. Roddick's company is valued at $1.2 billion; profits have risen 40 to 50 percent each year since the business started. In 1991 alone, 150 new shops were opened [2].

Return

Many organizations are making remarkable improvements in their environmental performance for reasons other than that these improvements reflect their core values or self-images. They are responding because they realize that they must comply with environmental regulations to stay in business and avoid costly fines. More and more organizations are recognizing that environmentalism has presented them with new opportunities to improve their operations and gain a competitive edge. The annual reports and other publications with environmental information from companies such as Dow Chemical, 3M, Northern Telecom, AT&T, and Chevron suggest that these organizations have been largely motivated by the *return* that they expect to gain from their measures to reduce the use of energy, reduce waste, recycle, reduce toxic emissions, and so on.

Expectation

Individuals and organizations will attempt what they have some hope of achieving. The perceived availability of *resources* and the perceived existence and strength of *blocks* are the primary determinants of how strong the expectation for success is.

Organizations that have a history of solving problems in creative ways and of quickly incorporating new technologies into their work processes will approach the environmental challenge with very different expectations from those organizations that have no such history. Organizations that have placed a high priority on their own capacities to learn and have accepted rapid change as the norm will not be intimidated by the environmental challenge but will view it as another challenge—something that they quickly learn how to manage.

One great disservice that government provides is to so burden businesses with specific regulatory, compliance, and reporting requirements that they despair of being able to respond and, consequently, develop low expectations. Government, to be a valued partner with business in helping the environment, must consider ways to empower business as well as to control business.

The responses that organizations are making to the environmental challenge cover a broad spectrum. In some organizations, there are strong, core values that energize the full and unqualified response to saving the environment. In other organizations, it is clear that "environment" is still a dirty word. These organizations are primarily concerned with staying just within the requirements of the various environmental laws and regulations.

In one organization that I visited, the director for management services had issued an order that "not one cent of this organization's budget will be spent on recycling programs." In another, TQM teams were explicitly prohibited from working on environmental problems.

The strength of the motivation to take positive action toward improving an organization's environmental performance will depend on the degree of congruence or *match* that exists between the organization's values and self-image. The strength of this motivation also will depend on the specific *return* that the organization expects to realize from its environmental initiatives. Both *match* and *return* interact with the strength of the organization's *expectation* to carry out its environmental tasks and reach its environmental objectives.

The Response Model for Sustainable Performance contains the following elements:

1. The pressures that are working on the organization to respond to the environmental challenge.

2. The screens that filter the information that leaders receive and use about the environment.
3. The motives that influence the responses that organizations make to the environmental challenge.
4. The necessary competencies (knowledge and skills) of the work force to be fully involved in SP.
5. The level of response to the environmental challenge that is being made by the organization.

We have discussed the first three of these; now we will focus on the fourth.

Competencies

Competencies affect the level of confident expectation that people have about being able to do anything. Competencies, therefore, determine in part the strength of motivation that people will have to work toward SP. However, even when people are strongly motivated to work toward SP, they still require special knowledge and skills to act fully on their motives, and it is only by continually upgrading the competencies of the work force that leaders can expect an organization to improve its response to the environmental challenge.

SP, like TQM, is a process of continuous improvement that is fueled by the mental abilities of the work force. The challenge facing organizations undertaking to transform themselves into ecological systems that are fully integrated with the natural environment is how to take full advantage of these mental abilities.

Words such as influence, involvement, and empowerment, which gained new popularity with TQM, have as their root the idea of using the full mental resources of the work force. TQM also brought new credibility to several good ideas that had been around for a long time, including the following:

- People closest to a job know best how to do it.
- People who do the work are the ones who can improve it.
- Problems should be solved at the lowest possible level in an organization.
- Teamwork is the foundation of quality.

The focus of training for SP is to equip people to exert competent influence over their jobs, their work groups, the processes in which they participate, and the total organization. In this regard, training for SP shares a great deal of purpose and content with TQM.

SP and TQM

In order for organizations to drive aggressively toward SP, they must already be fully committed to TQM and continuous improvement. SP is the next step in an organization's evolution toward total quality. SP can be achieved only when an organization

- Has the unquestioned commitment and support of management for continuous improvement.
- Focuses on processes and systems to improve its performance.
- Is driven by the needs of its customers (internal and external).
- Develops fully the mental resources of the work force and extends employee influence.
- Uses team formation and development as its primary organizing and improvement strategy.

- Uses measurement at all levels to determine where it has been, where it is, where it is going, and how it will get there [4].

The list of knowledge and skills described below for SP assumes that training in the knowledge and skills required for TQM is already in place or is being put into place. TQM training in team development, work-process improvement, customer satisfaction, measurement, rational problem solving, and statistical process control is also required for SP. In addition, training for SP includes the knowledge and skills that permit each person to work for the organization and for the environment. These competencies can be derived from the Response Model for Sustainable Performance and its various elements.

Knowledge and Skills for SP

The competencies needed by each member of the work force parallel those needed by all leaders. In SP as well as in TQM, every employee is a problem solver and a decision maker. In training people for SP, special emphasis should be placed on ensuring that they

1. Understand the milestones that their organization must reach in its progress toward SP.
2. Understand the pressures that are forcing organizations toward sustainable performance.
3. Have a clear understanding of the meaning of SP, its characteristics and principles.
4. Understand the level of response that the organization is making to the environmental challenge.
5. Are able to influence the level of response their organization is making and lead it toward SP.

6. Understand and use the general strategies and special tools for SP.

The milestones of SP, the general strategies for SP, and the tools for SP will be discussed in following chapters. It is not, however, the specific strategies and tools of SP that should be addressed first in SP training. The first and most fundamental training requirement is for employees to understand what SP is and why it is a critical issue for their organization's survival. The Sustainable-Performance Management Model (Figure 1-1), the Systems Model for Sustainable Performance (Figure 4-1), and the Response Model for Sustainable Performance (Figure 6-1) provide good overviews of the key issues and elements in SP.

Individuals and teams must understand the forces driving the organization toward SP, and they must have at least a basic understanding of the ecological issues that underlie these forces. SP requires that people respond to problems and opportunities in ways that are integrated and systemic, not fragmented and piecemeal. People must know how the pieces fit. These three models provide this kind of information.

Levels of Response

The Response Model for Sustainable Performance (Figure 6-1) assists leaders in managing SP by describing how and why organizations respond to the environmental challenge. The elements in the model provide an overview of how the interactions of pressures, screens, motives, and competencies influence the levels of response that organizations can make to the environmental challenge.

In addition, an understanding of response levels can be turned into a useful tool for assessing levels of response and for establishing baselines from which progress toward

SP can be tracked and improved. This process will described in a later chapter.

Results from a survey of the literature, reviews of what organizations report about themselves, case studies, and interviews with leaders in the field indicate that organizations usually begin with efforts to comply with environmental laws and regulations and then begin to move along a continuum that reaches toward the goal of SP. The major levels of response are (beginning at the lowest) as follows:

1. Compliance with the law
2. Nonintegrated initiatives
3. Integrated environmental plans and initiatives
4. Sustainable performance

Each of these levels has its own special characteristics. The level of response depends on the degree to which the ten Principles of Sustainable Performance have become manifest in the day-to-day life of an organization. An organization will move from simple compliance to full-fledged SP according to the degree to which

1. It is managed as an open system, and all decisions and key actions are undertaken with a view of the organization as a set of interdependent and interrelated parts.
2. All processes, products, and services are designed or revised to make them compatible with nature's ecosystems.
3. It has made the continuous improvement of results in sustainable development a priority.
4. The organization's stakeholders have been fully involved in its planning for SP.

5. The organization has developed accounting procedures for quantifying what it uses from the environment and what it returns to the environment.
6. The organization's planned and actual environmental performance is fully and openly communicated to all stakeholders.
7. Every member of the work force is involved in undertaking the plans and programs for SP.
8. Integrated processes for auditing, measuring, and reporting environmental performance have been developed and implemented.
9. Partnerships have been developed with entities such as government, business and industry, educational institutions, research and development groups, suppliers, and customers to discover and implement ways to improve SP.
10. All planning, decision making, and human resource systems have been made fully congruent with the organization's commitment to SP.

Compliance

The first stage of responsiveness is largely concerned with meeting the demands of law and regulatory requirements. At this stage, the organization has not become sensitive to many of the other pressures created by the needs of stakeholders, an aroused citizenry, green consumers, and the threat posed by greener competitors.

At this level, the organization begins to define its current level of compliance with environmental laws and regulations and its level of risk. Its goal is minimum compliance with minimum risk at minimum cost.

There are no proactive environmental policies, specific improvement plans, or goals. Organizations at this stage selectively use audits, establish tracking and reporting requirements, develop budgets for compliance costs, and publish formal responsibilities for managers and other key personnel to ensure compliance with regulations. There rarely are any environmental initiatives that are not defensive and focused on keeping the organization and its executives out of trouble. Problems with compliance are identified, and remedial actions that cannot be avoided are undertaken. At this stage, there also may be a few "environmental" programs that are not integral to the organization's work processes, services, or products, such as car pools, paper recycling, and eliminating the use of styrofoam cups.

Nonintegrated Initiatives

In this second stage, organizations begin to go beyond the strict requirements of the law to undertake various environmental programs, such as beginning to anticipate potential areas of liability, reducing the use of energy, taking advantage of obvious opportunities to reduce waste and packaging, and emphasizing conservation and good housekeeping.

Organizations in this stage begin to anticipate new developments in environmental regulations. Audits are used, but the focus is still largely on compliance. There is increased sensitivity toward the needs of stakeholders, and some organizations will formulate environmental policies.

The early evidences of SP have begun to emerge. There may be attempts to formulate measurable improvement goals, and specific projects may be underway to modify processes, services, and products to reclaim waste and to reduce pollution and emissions. Moves toward total quality environmental management through the use of tools such as life-cycle planning do not usually exist. Strategies to involve the total membership of the organization do not exist or are just emerging.

Integrated Environmental Plan and Initiatives

This is the first level of response at which the organization begins to position itself for achieving SP. In this first stage of development, most of the following initiatives have been started or planned:

1. Policy for sustainable performance and specific improvement goals published.
2. Baselines for environmental performance established, and improvement monitored against these baselines.
3. SP training underway for all members of the work force.
4. Specific improvement objectives and projects in place.
5. Investment made to develop or acquire new environmental technologies.
6. Ongoing auditing and reporting systems developed and implemented.
7. Information-sharing and problem-solving coalitions and partnerships built with industry members, stakeholders, governmental agencies, and professional groups.
8. Planning, decision-making, and human resource systems revised to support SP.

Sustainable Performance

The stage of sustainable performance is not a fixed condition. Like the process of total quality management, its dominant characteristic is continuous improvement. At this stage, the concepts of environmental responsibility and SP are fully embedded in the organization's information, planning, decision-making, and human resource systems. Sustainability is incorporated into the processes of design and production. The organization has grown its own environmental heroes and has its own environmental success stories.

It has improved its competitive position and demonstrated that it can "manage green." It has begun to show significant improvement to its bottom line by reduction, recycling, and reuse. It may have begun to shift to alternative energy sources. It has found new and profitable green niches in the market and is perceived as a source of managerial excellence for sustainable development.

SP defines the kind of changes that organizations must make to survive and it describes the goal toward which successful organizations will strive. SP defines the values and characteristics of organizations that will become the norms for doing business in the future.

Organizations that have begun to function in accord with the Principles of Sustainable Performance have developed the full involvement of their stakeholders and have implemented initiatives such as community panels; use of environmental experts and advocates in organizational management and governance; and input from stakeholders, customers, and suppliers. They have developed ways to maintain open communication with their stakeholders through annual reports, special environmental reports for internal and external consumption, special feedback and dialogue sessions, tours, and news conferences. They have developed the full involvement of employees in every aspect of their SP initiatives by aggressive use of the best TQM practices.

We can expect an organization that has begun to operate within the principles of SP to be operating as follows:

1. Publishes a policy for sustainable performance and specific improvement goals with the full involvement of all its stakeholders and has a plan for revising this policy and these goals to reflect new and more rigorous improvement initiatives.
2. Develops and revises baselines for environmental performance for all key inputs from the environment and all key outputs into the environment (direct and indirect). Maintains data on all processes to reclaim waste and secondary products.
3. Has a well-developed SP training program. Ensures that training for SP is fully integrated into its overall training program; that training is provided in new environmental technologies; and that quality time is provided to support and stimulate learning, acquisition of new technologies, and innovation for SP.
4. Has improvement objectives and projects in place. Creates specific projects to design or revise all services and products to ensure their full compatibility with nature's ecosystems (e.g., projects to reduce packaging, employ closed-loop processing cycles, or market waste as a resource for other companies).
5. Has implemented processes that support research and the development of new technologies and uses these to improve its environmental performance.
6. To assess and track environmental performance, uses a fully developed and integrated auditing system that includes not only information on technical performance such as compliance, emissions, and waste, but also information on management

performance and the performance of all support systems.

7. Initiates and participates in coalitions and partnerships with industry members, governmental agencies, and professional groups to share information, do joint problem solving, develop standards, and support the development of technology.
8. Has made the environment a part of all management systems, so that environmental performance is part of the organization's information-management system and information about environmental performance is reported with other key performance data.
9. SP is understood by all members of the organization to be a job responsibility, and all human resource systems have been revised to support SP.

Chapter 7 contains a *Sustainable-Performance Assessment* survey that can be used to (1) develop a baseline for an organization's response to the environmental challenge; (2) determine at what level the organization is responding; and (3) design strategies for improving the level of response.

Summary

The Response Model for Sustainable Performance suggests leadership tasks that must be performed in managing the milestones and helping the organization move toward SP. The Sustainable-Performance Management Model (Figure 1-1) suggests the competencies required to get there. Five sources of information can inform the management decisions required for building SP into an organization. These sources include

- The pressures working on the organization.
- The characteristics of sustainable performance.
- The Systems Model for Sustainable Performance.
- The Principles of Sustainable Performance.
- The level of response that the organization is able to make to the environmental challenge.

The competencies that leaders need in order to help their organizations respond fully to the environmental challenge are

1. Understanding the milestones that their organizations must reach in their progress toward SP.
2. Understanding the pressures that are forcing organizations toward sustainable performance.
3. Understanding the meaning of SP, its characteristics and principles.
4. Understanding the levels of response that organizations can be expected to make to the environmental challenge.
5. Knowing what level of response their organization is making.
6. Being able to influence the level of response their organization is making.
7. Understanding and using the general strategies and special tools for SP.

Thus far, I have provided the content for developing the first six competencies. The next chapter will present the competencies associated with the strategies and tools of SP. Understanding and using these strategies and tools is critical to the process of managing SP.

References

1. Dillon, P., & Fischer, K. (1992). *Environmental Management in Corporations: Methods and Motivations.* Medford, MA: Tufts University.
2. Interview with Anita Roddick. (October, 1992). *Business Ethics,* pp. 27-29.
3. Kiechel, W. (February 4, 1989). How Executives Think. *Fortune,* pp. 102-118.
4. Kinlaw, D. (1992). *Continuous Improvement and Measurement for Total Quality: A Team-Based Approach.* San Diego, CA: Pfeiffer & Company.
5. Kinlaw, D. (1991). *Motivation Assessment Inventory.* Norfolk, VA: Kinlaw Associates.

7

STRATEGIES AND TOOLS

The Sustainable-Performance Management Model (Figure 1-1) shows a series of milestones surrounded by sources of information for managing these milestones on the way to SP. In this chapter, I will discuss the following sources of information from which knowledge and skills can be developed:

- General strategies for improving SP
- The Sustainable-Performance Assessment
- Auditing
- Benchmarking
- Life-Cycle Analysis

In the model, the Sustainable-Performance Assessment, auditing, and benchmarking are sources of information for managing the milestones, but because each of these is also a tool, they are described in this chapter as tools.

The competencies that leaders must have to manage SP are as follows:

1. Being able to lead their organizations through the milestones that must be reached in the movement toward SP.
2. Possessing a clear understanding of the pressures forcing organizations to move toward SP.
3. Having a clear understanding of the meaning of SP, its characteristics and principles

4. Understanding the levels of response that organizations can be expected to make to the environmental challenge.
5. Knowing what level of response their own organization is making.
6. Being able to influence the level of response their organization is making and lead it toward SP.
7. Understanding and using the general strategies and special tools for SP.

In the previous chapters, I have described the first six of these competencies. In this chapter, I will discuss the seventh competency.

This chapter has two sections. The first describes the general strategies that are producing payoffs in environmental performance. The second presents the tools for SP.

Section I: The General Strategies

A number of strategies have proved to be valuable in moving organizations toward SP. There are five general strategies that are clearly compatible with the principles of SP; they lead to profit and responsible environmental performance. Most often, these strategies do not exist in a discrete way. We often find them in combinations, and a single initiative may include several of the strategies. The five strategies are

1. Practicing conservation and paying attention to every detail associated with a work process, e.g., using only the necessary amounts of materials for a process, turning off the water, turning off the lights, keeping machinery and vehicles in top running condition.

2. Modifying or replacing existing processes, products, and services to make them environmentally friendly, e.g., changing to more energy-efficient machinery, reducing packaging materials, eliminating toxic chemicals and emissions.
3. Reclaiming, by recycling and reusing, waste and secondary products such as chemicals, paper, plastic, metal, and water.
4. Reducing the use of materials, e.g., reducing the amount of packaging or packing used, reducing the size of reports and invoices, reducing the amount of a material used in a process, and reducing energy use.
5. Finding new, "green" niches in the marketplace and delivering new services and products, e.g., waste management and disposal, alternative energy sources, alternatives to toxic and ozone-depleting chemicals, and benign cleaning agents.

These strategies are not always completely distinguishable from one another in application. Improvements often employ one or more strategies at the same time. It is useful, however, to isolate each of the strategies in order to further our understanding of them and to build on them to develop new strategies.

Strategy One: Practicing Conservation and Paying Attention to Details

Most work processes can be undertaken in a more environmentally sensitive way just by paying attention to how energy is used, how materials are consumed, and what wastes are produced. Practicing conservation and attending to details means doing obvious, common-sense things such as keeping equipment in top working condition, minimizing spills, eliminating rework, and reducing scrap. All such actions are integral to doing a quality job. *Doing a quality job with whatever equipment and resources are at hand is an immediate and accessible way to move toward SP. Inefficiency, in whatever form it may take, is the antithesis of SP.*

Cleaner production often requires no new cash investment and can be accomplished with existing facilities. What is required is management attention and the involvement of the work force. Just by taking a new and creative interest in how water, energy, and materials are used, people can create improvements in things such as wastestream separation, better monitoring of emissions and processes, waste recycling and disposal, and tightening of supplier requirements.

The United States Office of Technology Assessment estimates that 50 percent of all environmentally harmful industrial wastes could be eliminated with the technology that was available in 1986 [10]. Much of this simply amounts to ensuring maximum quality and minimum waste in any process.

This strategy is useful in developed countries as well in developing ones. A textile mill in Bombay, India, increased the collection rate of caustic soda from its wash waters from 75 percent to 85 percent and the recovery rate from 81 percent to 90 percent just by correcting leakages and seepage, paying more careful attention to the washing process, and improving filtration. These good-housekeeping procedures resulted in large savings of caustic soda per day and a savings to the company of $500,000 per year [1].

Using the minimum amount of energy, raw materials, and time always has been considered a good performance practice. It is reemphasized with the advent of TQM and must be noted again as fundamental to good environmental performance.

Strategy Two: Modifying or Replacing Existing Processes, Products, and Services

From input, through manufacturing and production processes, to output, every process and the products and services that result from processes have the potential for improvement. From the perspective of SP, improvement means reducing the negative impact on the environment.

The Chevron Oil Company eliminated the use of toxic chemicals in one of its cleaning systems with an aqueous process. It replaced the system with a closed-loop, hot-water cleaning system, which resulted in a $50,000 annual savings by avoiding the previous costs of chemicals and chemical-waste disposal.

General Dynamics replaced a chromic-acid, aluminum-anodizing system with a new, computerized sulfuric-acid anodizing system that utilizes computerized hoists and on-demand rinsing. The new system produced the following results:

- A major reduction of chromium emissions.
- Reduction of rinse-water requirements from twenty gallons per minute to eight gallons per minute.
- A cost savings for electricity of $11,000 per year.
- Reduction of wastewater-treatment costs due to decrease in metals-removal requirements.
- Reduced labor costs [13].

In Shreveport, Louisiana, AT&T replaced a cleaning solvent that contained CFCs with soap and water. CFC use dropped from 146,000 pounds per year to 10,000 pounds in a period of three years. In Richmond, Virginia, AT&T achieved an even more dramatic drop in the use of CFCs by replacing solvent-based cleaners with water-based cleaners. CFC use was reduced from three million pounds of chlorinated solvent emissions to less than 1,000 pounds between 1987 and 1990 [14].

Strategy Three: Reclaiming Waste and Secondary Products

The words "recycle," "reuse," and "reclaim" are used to describe actions and processes by which waste, scrap, and whole production elements are fed back into one system or into another as input. There are no established meanings of these words. For the sake of clarity, I will use "reclaim" to describe any action or process that leads to the reuse of a material or production element.

The Ford Motor Company of Germany announced in 1992 that its new supplier, the Nauerz Company, will provide disassembly and disposal services to Ford owners in the Cologne area.

The Medical Center of the University of California at Irvine in Orange, California, has cooperated with California Edison in an energy-efficiency program that now saves the hospital $400,000 annually in energy costs [2]. The medical center also has installed a system called microwave disinfection, which shreds and steams intravenous waste in a microwave chamber, achieving an 80 percent reduction in volume. The savings from landfill and incineration costs is estimated at $18,500 annually.

One of the most innovative projects for reclaiming waste is Anheuser-Busch's program to use fuel generated from brewery wastewater to power its plants' boilers. This bioenergy-recovery system will produce 500,000 cubic feet of methane gas per day at its Baldwinsville plant (the equivalent of 2,250 gallons of oil). By producing its own power, the plant will save ten million kilowatts of electricity each year and will save the company $1 million per year in the costs of gas and electricity combined.

In the United States, landfill costs for hazardous waste went from around $80 per ton in 1980 to around $255 per ton in 1990 [5]. In Europe, waste processing can now cost organizations anywhere from $380 per ton for solid waste to $10,000 per ton for toxic and hazardous wastes [12]. Introducing new recycling processes is becoming more and more attractive to organizations as a strategy to achieve SP.

The Henry Ford Hospital in Detroit, Michigan, is working with other hospitals to recycle plastic materials from intravenous (IV) systems. Hospital workers deposit used IV materials in a designated container (no sorting needed). The materials are collected and sent to a processor for sorting, cleaning, and use in new products such as vinyl floor runners and mats.

AT&T saved $1.4 million by recycling paper in 1984. In 1988, dumping fees were $156,000 per year in New Jersey. Today, recyclers are paying IBM's Bedminster offices $10,500 per year to take paper waste off its hands. In addition to the revenue, IBM has halved its dumping cost.

In the recent past, Dow Chemical was spending $10 million per year to neutralize and flush away dirty hydrochloric acid that was left over from making commodity chemicals such as chlorine, polyethylene, and caustic soda. It paid an additional $10 million per year to buy new acid. A few years ago, Dow formed its Gulf Coast Acid Team, a task force of employees asked to present a better idea. The result is that, today, a network of barges collects the spent acid at locations in Texas and Louisiana, purifies it, and transports 99 percent of it back for reuse in the manufacturing system. The payoffs are $20 million savings per annum and cleaner water.

Monsanto's Muscatine, Iowa, plant makes ABS, a plastic used in automobile parts and household appliances. The managers of the plant recognized that state and federal air-quality laws could wreak havoc with their production system. They installed a system for capturing and reusing chemical vapors that escaped from the process. The new system cost $10 million but it will save $1.5 million per year in labor, energy, and cost of materials [3].

U.S. Industries now spends as much as $80 billion per year in end-of-pipe pollution-control technology and equipment. It makes economic sense to prevent pollution in the first place.

One of the many companies that is cashing in on pollution control is Riker Laboratories in California. The company is currently saving $15,000 annually since it replaced a chemical solvent with a water-based solvent for coating medicine tablets. It also realized a one-time cost savings of $180,000 for pollution-control equipment that was deemed unnecessary because of the change.

Ciba-Geigy invested $300,000 in process changes and recovery equipment at its Tom River plant in New Jersey and reduced disposal costs by $1.8 million between 1985 and 1988 [6].

Also at Ciba-Geigy, the output of useable products represented only 30 percent of the company's total output in 1979. The rest was waste, for which the company had to carry very heavy management and disposal costs. By 1988, Ciba-Geigy had increased its efficiency ratio of product to waste by over 60 percent. In one process alone, it was producing a metric ton of a chemical called amide, for which three tons of trichloride and twelve tons of water were required. The process produced fourteen tons of effluent. The old system has now been replaced by a new one that uses only 1.9 tons of raw material per ton of produced amide. In addition, the new process uses no water and only 0.6 tons of ascetic acid, which is recycled into other processes.

The reuse of chemical or constituent materials in a process is not rare. What is rare is the reuse of whole product elements. Xerox Corporation has demonstrated a very innovative process of reuse. It is now producing copiers made of only 20 percent new components for its European market. The remaining 80 percent of components have been repaired, reconditioned, and reassembled into the new machines.

In 1988, Coca-Cola Company formalized a recycling program called "Recycling Works" for its own used cans with Alcoa Aluminum. Alcoa also has developed ties with United Airlines and leases a forty-foot box trailer to the United group at Newark to collect its used aluminum cans. Now, rather than paying $100 per ton in hauling fees in Newark, United sells its aluminum cans for about $25,000 per year [9].

Strategy Four: Reducing the Use of Materials

Reduction of the use of energy or materials can occur at any point in production process, from input to the packaged product. Some of the easiest gains occur in reducing packaging.

Through its energy-management programs, IBM has saved $32 million for 1990 and 1991. IBM estimates that the unused 340 million kilowatt-hours cut carbon-dioxide production associated with the generation of electric power by more than 210,000 tons. That is equivalent to taking 50,000 automobiles off the road.

IBM also has distributed guidelines for specific energy-management practices to its major facilities worldwide, and these sites are now subject to regular energy-efficiency audits [5].

Hewlett-Packard (HP) has undertaken a variety of environmentally useful initiatives. It has changed its shipping cartons from bleached boxes to unbleached kraft boxes, because the production of the latter uses less energy and produces fewer toxic emissions. It is also using recycled polystyrene foam in some of its packaging. One significant reduction in the amount of packaging that it uses occurs in shipping its Vectra computer. It has reduced the number of boxes from three to two. This simple reduction is estimated to save HP $2 million per year in packaging costs.

Another way to reduce the amount of materials used in production (and therefore the amount of waste downstream) is to make things last longer. Patagonia, a manufacturer of outdoor-activity and sports clothing, has determined that clothing products are as much of a pollutant as automobiles, because cotton is produced with pesticides, defoliants, and formaldehyde. Wool is produced from sheep, which are notorious destructive grazers. Patagonia's approach is to serve the environment by making quality goods that last longer [5].

Body Shops International, which I have already cited for its environmentally sensitive performance, encourages the return and reuse of the bottles that contain its lotions and toiletries by paying customers $.05 for each bottle returned. Each Body Shop also offers refillable containers and continues to reduce the amount of packaging used with orders.

Reduction of materials can occur in small ways that add up to big savings over time. AT&T permits its customers to choose a compressed version of their monthly bill, which requires less paper than the standard one.

Sears, Roebuck and Company, in cooperation with its suppliers, has begun a packaging-reduction program that will reduce the volume of packaging materials for products sold at its stores by 25 percent by the end of 1994. Sears estimates that the program will reduce packaging by 1.5 million tons by the end of 1994 and will save the organization $5 million per year.

We have now identified four general strategies for improving an organization's environmental performance: conservation and attending to details, modifying and replacing, reclaiming, and reducing. The fifth strategy, finding new markets, will gain in popularity as we proceed into the environmental age.

Strategy Five: Finding New, Green Market Niches

A fifth strategy for SP is to find new markets for old or new materials (such as waste products from a process) or to develop new services and products to meet the expanding need for such services and products.

I am not concerned in this book with the whole new environmental industry that has developed—especially the industry concerned with waste management and disposal. In this section, I will focus on organizations that are not primarily in the environmental business but which are finding that they can market their wastes or special services or can create new products congruent with their own capabilities and missions.

ARCO's Los Angeles plant has introduced a complex series of changes that have reduced waste volumes from 12,000 tons to 3,400 tons per year and have generated a revenue of $2 million per year in disposal costs. The plant sells its spent alumina catalysts to Allied Chemical and sells its spent silica catalysts to cement makers. Previously, these materials were classified as wastes and disposed of in landfills at a cost of $300 per ton.

Data General has a goal to eliminate landfill waste. One major component of its program has been to find markets for the spent chemicals and sludges that come from its production processes. The organization was able to find potential buyers and to modify its production and waste-treatment processes to produce the kinds of wastes that its buyers would accept [13].

Del Monte's pineapple plant in the Philippines sells its pineapple waste for use as cattle feed for $50,000 per year. It avoids additional hauling costs of $55,000, for a total benefit of $105,000 per year.

At its Wilmington, North Carolina, division, General Electric entered into a joint venture with Federal Paper Board for that company to use 4.5 million gallons per year of ammonium nitrate waste as a nutrient feed for Federal's waste treatment plant.

Fifteen California organizations have formed the Warner Center Association Recycling Program. The program is a closed-loop recycling venture. Participating companies will collect and recycle their own waste materials, which amount to 3,300 tons per year. Weyerhaeuser Corporation will reprocess the materials into new products, and Royal Supply will sell them to member companies.

British Petroleum America recovers and sells acetonitrile (ACN) to the pharmaceutical industry. American Cyanamid has undertaken a similar project at its Fortier plant in Westwego, Louisiana. This project also will recover cyanide for reuse in the production of chemicals for making clear plastic [3].

Summary of Section I

Five general strategies currently are being used by organizations to respond to the environmental challenge. They are as follows:

1. Practicing conservation and paying attention to details associated with work processes.
2. Modifying or replacing existing processes, products, and services.
3. Reclaiming secondary products such as chemicals, paper, plastic, metal, and water.
4. Reducing the use of materials.
5. Finding new, "green" niches in the marketplace.

Knowledge of key strategies, like knowledge about the principles and characteristics of SP, is required of leaders who will manage their organization's performance in the environmental age. They must know the milestones that their organization must reach in the journey toward SP. Leaders also need to understand the pressures to which organizations must respond and the levels of response that their organizations are making to the environmental challenge.

One additional competency that leaders need is knowledge of the tools that are essential elements in any plan for SP. These tools are described in the next section.

SECTION II: TOOLS FOR SP

The following tools will be described in this section:

- The Sustainable-Performance Assessment
- Auditing
- Benchmarking
- Life-Cycle Analysis

The Sustainable-Performance Assessment (SPA)

The Sustainable-Performance Assessment is included in the Appendix. This section introduces the SPA and suggests how it can be used as a tool for improving SP.

The SPA is a survey instrument. It can be used to develop baselines for tracking and monitoring an organization's overall response to the environmental challenge. More importantly, it can be used to identify improvement opportunities.

The SPA is based on the description of SP developed in this book. It uses the milestones in the Sustainable-Performance Management Model. It also uses the three qualitative characteristics required to manage the milestones. These characteristics are

- The demonstrated commitment of management to SP.
- Determination to involve the work force.
- Determination to involve the organization's other stakeholders.

The SPA measures the level of response that an organization has reached (see Chapter 6). These levels are

1. Compliance with the law.
2. Nonintegrated initiatives.
3. Integrated environmental plan and initiatives.
4. Sustainable performance.

The SPA also includes a "no response" level to identify areas of complete inactivity.

An organization's performance in reaching each of the milestones is measured by using seven criteria. For example, the criteria for reaching the milestone of publishing a policy are as follows:

- Management demonstrates full support.
- Policy developed with full involvement of work force.
- Policy developed with full involvement of all stakeholders.
- Policy fully compatible with the Principles of SP.
- Policy leaves no doubt about organization's valuing of the environment.
- Policy ties business interests of organization to best interests of environment.
- Policy specifically includes the concept of sustainability.

The SPA can be used in the following ways:

1. As an organizational-assessment tool. In this case, the whole work force (or an appropriate sample) completes the SPA, and the data is used to develop a baseline of the organization's response level.
2. As a management-feedback tool. In this case, managers use the SPA to communicate their own perceptions of how the organization is responding to the environmental challenge.
3. As a third-party audit tool to evaluate the organization's level of environmental performance. In this case, the audit team uses the SPA and, from interviews, analysis of documents, direct observation, etc., provides an assessment of the organization's performance.

The reason for using the SPA is to provide management with information that it can use to improve the total organization's response to the environmental challenge. Such a response is the responsibility of management. Information from the SPA should be processed by management as it addresses the following questions:

1. How completely and unambiguously have we demonstrated commitment to SP? What must we do to demonstrate our commitment more fully?
2. To what degree have we developed a work force that is fully involved in SP? What must we do to increase this involvement?
3. To what degree have we involved the organization's other stakeholders in its move to SP? What must we do to increase this involvement?
4. How is our performance on each of the eight milestones measured on the SPA and what can we do to improve each?

The key to the successful use of the SPA, as with any other survey tool, is how well the organization prepares itself to use the information. Here are a few suggestions:

1. Ensure that management is fully involved in planning to use the SPA and has a good understanding of what the SPA measures.
2. Put in place a specific plan for using the information from the SPA. This is a feedback plan and should include at least the following:
   - Clarity about how the information will be summarized and presented.
   - Clarity about who will be responsible for doing what with the information obtained.
   - A schedule for following up the action taken on the information obtained.
   - A plan to involve the work force and other stakeholders in using the information.

- A plan to communicate fully and regularly to the work force and other stakeholders about how the information from the SPA is being used.

The SPA is a tool for developing baselines and identifying opportunities for improvement. Another tool for developing baselines is the environmental audit.

Audits

Environmental auditing originated as a way of determining an organization's degree of compliance with regulatory requirements. It has become a more proactive tool for developing environmental-performance baselines that can be used for total quality environmental management.

There are many types and sizes of audits. The term audit sometimes is used interchangeably with "assessment" and "risk assessment." One organization now makes a distinction between "environmental" auditing and "ecological" auditing to emphasize the difference between the traditional, defensive, compliance approach and the more proactive approach [4]. To ensure that the audit is used as a practical tool for improving environmental performance, it is useful to make a few distinctions and to develop concrete definitions. First of all, it is necessary to distinguish between audits and auditing programs.

Auditing Programs

The purpose of an auditing program is to produce comprehensive information about the organization's total environmental performance. An auditing program includes any number of audits. These audits collect data on all the key variables that determine an organization's level of SP.

The framework for understanding the scope of an auditing program is suggested by the Systems Model for Sustainable Performance (Figure 4-1). A fully developed, environmental auditing program will produce a comprehensive description of what the organization has done, what it is currently doing, and what it needs to start doing. Such a program will include audits that assess every significant organizational input from the environment and every output to the environment.

Audits are tools for measuring current levels of SP, which is the evolution of organizations into wealth-producing systems that are fully compatible with the natural ecosystems that generate and preserve life. A system may be a facility, an organizational unit, a work process, an underground storage tank, and so on. Audits, then, help to determine just how compatible with the natural ecosystems every aspect of an organization's performance is.

> An audit is a systematic collection of data that can be used to determine the full impact of a system's input from the environment or output to the environment.
>
> An auditing program is a systematic collection of comprehensive data that can be used to determine the full impact of an organization's input from the environment or output to the environment.

Developing an Auditing Program

Developing an auditing program is one of the milestones for sustainable performance. If senior management has not already committed to SP, published a policy for SP, and moved through the other milestones (see Chapter 2), it is premature to try to develop an integrated and comprehensive auditing program. Management will not be committed to an auditing program that is focused on continuous environmental improvement unless it is, first of all, committed to SP.

The process of developing an auditing program is parallel to that of developing a measurement program. It entails a set of steps similar to the following:

1. Ensuring the commitment of senior management to the rationale for and payoffs of an auditing program.
2. Developing the questions that senior management wants to answer.
3. Developing reporting requirements and process.
4. Ensuring the regular review and revision of the entire auditing program.

Step One: Ensuring Commitment of Senior Management to the Rationale for and Payoffs of an Auditing Program

This includes identifying the expected payoffs, describing the level of effort required and the scope of the program, and identifying senior management's responsibilities and operational responsibilities. The International Chamber of Commerce has identified the potential payoffs as [7]:

- Facilitating comparison and exchange of information between operations and plants.
- Increasing employee awareness of environmental policies and responsibilities.
- Identifying potential cost savings, including those resulting from waste minimization.
- Evaluating training programs and providing data to assist in training personnel.

- Providing an information base for use in emergencies and evaluating the effectiveness of emergency response arrangements.
- Assuring an accurate, up-to-date, environmental data base for internal-management awareness and decision making in relation to plant modifications, new plans, and so on.
- Enabling management to give credit for good environmental performance.
- Helping to assist relations with authorities by convincing them that complete and effective audits are being undertaken and by informing them of the type of procedure adopted.
- Facilitating the obtaining of insurance coverage for environmental-impairment liability.

Some additional payoffs that a well-designed and well-managed audit program can be expected to produce are

- Demonstrating in a very concrete way an organization's commitment to environmental protection to its employees and all its other stakeholders.
- Verifying the organization's compliance with environmental regulations.
- Reducing exposure to litigation, incidents, and adverse publicity.
- Assisting the organization to be a full participant in the exchange of environmental performance data by which all can benefit.
- Assisting in the exchange of information among all organizational components, factories, facilities, and so on.

The payoffs from an auditing program obviously will depend on the kinds of audits in the program, and these will depend on the nature of the organization's business and the kinds of questions that decision makers want to answer.

The second step in designing an auditing program is to determine the kinds of questions to be answered.

Step Two: Developing the Questions That Senior Management Wants to Answer

An audit, like any form of measurement, has minimal utility unless it is founded on a question that must be answered in order to ensure the continued success of the organization. Of course, the questions that are important at the initiation of an auditing program can be expected to change. An auditing program is an evolving process.

Some questions must be answered. These are the ones that are related to compliance with regulatory requirements. However, auditing programs that support SP are tools for continuous improvement and will go far beyond questions related only to compliance.

Auditing programs must be developed with the full involvement of the work force and the organization's other stakeholders. The kinds of questions to be asked, therefore, should be developed with the involvement of stakeholders. (Involvement as a quality required in managing all the milestones of SP is discussed in Chapter Two.)

When audits are understood as tools for SP, auditing programs will be designed around the following kinds of questions:

- What is our energy use?
- What are the breakdowns of the ultimate sources of energy we use, i.e., coal, oil, gas, nuclear?

- What are the amounts—by volume and weight—of the materials we use?
- What wastes—by kind, volume, and weight—do we produce?
- What ground-water contaminants do we produce?
- What emissions do we produce at known exit points—by volume and particle count?
- What fugitive emissions—by volume and particle count—do we produce?
- What wastes—by kind, volume, and weight—do we produce through our packaging?

The design of the auditing program also may include questions about the reclamation and reuse of secondary products and wastes, questions about the level and adequacy of training for SP, and questions about supplier performance and the life cycles of products and machines.

To answer a question related to SP, data must be retrieved by tracking the performance of the appropriate variables. A variable is any source of information that can help to answer the question being asked.

For example, if the question is "What wastes—by kind, volume, and weight—do we produce through our packaging?," the variables that could produce answers might include the following:

- Packaging by type (e.g., wood, plastic, paper, cardboard) for each product delivered.
- Percentage of packaging by kind currently being recycled.
- Total percentage of packaging by kind that could be recycled.
- Amount of energy associated with producing the packaging.

- Transportation costs associated with the acquisition of packaging.
- Transportation costs associated with packaging in the delivery of products.

The identification of the specific variables to be tracked and the design of a data-collection process to track these variable are elements in any specific audit. It is the responsibility of senior management to identify the questions that should be answered and the variables that must be tracked to answer these questions, although senior management may not necessarily be directly involved in the actual process of data collection.

The first two steps in designing an auditing program are ensuring that senior management is committed to such a program and identifying the questions that the program should answer. The third step is to decide how the data, conclusions, and recommendations that are developed through the auditing program will be communicated.

Step Three: Developing Reporting Requirements and Process

Key elements in an auditing program are the reporting requirements and process. The basic questions to be answered are: What information will be reported? To whom will the information be reported? How often will the information be reported? How will the information be used?

How reports will be made and used will depend on the kind of environmental policy the organization has and the kind of payoffs it expects to receive. If audits are viewed as tools for the continuous improvement of SP (to improve performance toward the environment and to improve the organization's competitive position), reporting information from audits will meet the following criteria:

- Provide information that requires action.
- Stimulate and support the exchange of information.
- Stimulate and encourage the involvement of the work force in improving the organization's SP.
- Stimulate and encourage the involvement of all other organizational stakeholders in improving the organization's level of SP.

Action. The first criterion that a reporting process must meet is that it delivers information that requires action. As with any measurement systems, if audits are not derived from specific questions and directed at achieving specific objectives, they can produce information that may lead to no improvement.

I have been involved on a number of occasions in assisting organizations to develop measures to help them track and improve performance. In the typical situation, the organization will already have started the process of developing measures and will have begun to experience a variety of problems, including the following:

- Grumbling and complaining from managers and members of the work force, saying that they don't see the purpose of measuring.
- Work units developing all kinds of measures that are used for display purposes but are not tied to any decisions.
- The perception that measuring is "busy work" and unrelated to the real business of "getting the job done."
- A lot of time wasted in trying to discover what to measure and how to measure.

- Reinforcement of resistances to measuring and the development of behaviors to avoid or even sabotage attempts to measure.
- A growing belief by managers that performance measurement will not ever become a highly useful management tool.
- An increase in the number of training programs to teach people how to measure.

The underlying causes for most of these and other problems related to measurement is that management has never developed for itself a clear rationale for measurement, has never identified the payoffs that it expects to receive from a measurement system, and has never identified the specific questions about performance that it wants to answer.

To drive action, the reporting process of an auditing system must be clearly tied to a set of questions that management wants to answer about its environmental performance and it must support the payoffs that managers want from the auditing program.

The first criterion that a reporting process must meet is that it drives action, i.e., that it requires decisions and leads to improvement initiatives. A second criterion that a reporting process should meet is that it stimulates and supports the exchange of information.

Exchange of Information. The sixth principle of SP (see Chapter 5) is

> Sustainable performance is an open process and requires that organizations communicate fully all aspects of their planned and actual environmental performance to all organizational stakeholders.

For an auditing program to conform to this principle, the reporting process must support the full exchange of information with all stakeholders including, of course, the work force. The expected payoffs from an auditing program that I have listed previously suggest just how important it is that the reporting system stimulate and support the sharing of information. None of these payoffs can be realized unless information about environmental performance is fully shared within the organization and with the organization's stakeholders. The auditing program is one of the key sources of information about environmental performance.

The full sharing of information obviously requires more than the publishing of reports. Like so many information systems in organizations, reports focus on what is being sent, not on what is being received. The sharing of information that leads to competent decisions and improvement initiatives is best done through a variety of interpersonal interchanges in groups and teams.

Reporting the information obtained through an auditing program should lead to action and to extensive sharing of information. The reporting process also should stimulate and encourage the involvement of the work force in improving the organization's level of SP.

Involvement of the Work Force. All the milestones of SP should be managed so that the full involvement of the entire work force is facilitated. Opportunities for such facilitation exist at every phase of an auditing program. The successful translation of an auditing program into initiatives to improve performance, as with all such initiatives, depends on the full involvement of the work force.

Involvement of people in the reporting process is created to the degree that

- The people most directly affected by any specific audit are fully involved in designing and conducting the audit.

- The people most directly affected by an audit are made responsible for using the results of the audit.
- The design for using the results of any audit is team-centered and uses audit-response teams.
- The audit-response teams represent all the needed competencies and the all the systems and organizational elements affected by the audit.
- The teams that receive information from the audit are responsible for making the improvements indicated by the audit.

The first three criteria for designing a reporting process are that the reporting process produces information in a way that leads to action, the information is fully shared, and the work force is involved in using the information. A final criterion is that the reporting process also fully involves all other stakeholders.

Involvement of Other Stakeholders. I have already stated that the reporting process of an auditing program, if it is to be congruent with a commitment to SP, must conform to the sixth principle of SP. Two additional principles are relevant to the criterion that the reporting process should stimulate the involvement of all other stakeholders.

Principle Two

Sustainable performance is an ecologically interdependent process and requires that all organizational processes, products, and services be revised or replaced to ensure their compatibility with nature's ecosystems.

Principle Four

Sustainable Performance is a community-building process. This means that organizations need to cooperate with one another and use the environment in ways that are equitable for one another. It also requires that organizations involve all their stakeholders in the processes of planning and implementing sustainable performance.

To be congruent with the principles of SP, the organization must develop a system of communication with all its stakeholders. One of the foundations for such a communication system is the organization's auditing system. A number of organizations have developed communication systems with their stakeholders that can serve as examples. Among these are IBM, 3M, Proctor & Gamble, Alcoa, Norsk Hydro, Hewlett/Packard, DuPont, and Chevron. Underlying all of these communication systems are well-developed auditing systems.

Audits provide the data for developing baselines. Audits also provide the data for monitoring performance and reporting changes in performance.

We have discussed the first three steps required of senior management in creating an auditing program.

1. Ensuring the commitment of senior management to the rationale for and payoffs of an auditing program.
2. Developing the questions that senior management wants to answer.
3. Developing reporting requirements and process

Step Four: Ensuring the Regular Review and Revision of the Entire Auditing Program

Auditing is a dynamic process that will change as progress in SP takes place, as new regulatory requirements are created, and as new opportunities for improvement are recognized. The final step for senior management in creating an auditing program is to ensure that the entire program is reviewed and revised to meet these new requirements and to take advantage of new opportunities.

One of the major reasons to revise an auditing program is the rapid change and increase in regulatory requirements. SP means going beyond compliance. To go beyond compliance obviously requires that all regulatory requirements are met.

In Chapter 3, I discussed the rapid changes in environmental legislation that are being generated at the local, state, and federal levels. It will be not only the pressures from regulatory requirements that will demand the regular revision of auditing programs, but all the other pressures that I have described (e.g., threat of personal culpability, environmental activist groups, an aroused citizenry, coalitions, international codes) as well. All these pressures will create new demands on the organization to provide new and different information about its environmental performance. The auditing program provides the data base from which information can be extracted and the source for the new information that may be required.

The review and revision of the organization's auditing program must become a standard management requirement and function. Such a review is an example of the kind of change that must take place in an organization's management systems in order for the organization to support SP.

Thus far, I have described auditing programs and the steps that senior management must take to develop such programs. Auditing programs are, obviously, made up of audits. Next, I will describe the environmental audit.

Environmental Audits

An audit is defined as

> A systematic collection of data that can be used to determine the full impact of a system's input from the environment or output to the environment.

Audits come in endless shapes and sizes. They can be narrowly focused to answer restricted questions such as "What are the emissions from a specific process?" or "What is the energy use of a specific set of machines?" An audit can be more broadly focused to evaluate a single plant or facility; it can be designed to assess the total environmental performance of a functional unit; or it can be designed to assess the performance of a total organization in terms of a particular set of variables.

Audits can be focused on functional units of an organization, such as the following:

- Health and medical
- Research and development
- Purchasing
- Engineering
- Production
- Marketing
- Sales
- Personnel
- Training and development
- Transportation
- Administration

If audits are focused on the environmental impact of outputs, they will assess waste, emissions, and pollution. The audit list proposed by the EPA can serve as an example. The areas proposed by the EPA include the following:

- Air
- Asbestos
- Drinking water
- Water pollution
- Nonhazardous waste
- Hazardous waste
- Underground storage tanks
- Past disposal of hazardous materials
- Emergency planning
- PCB (polychlorinated biphenyls) management
- Pesticides
- Radioactive materials
- Environmental noise

The Appendix contains a selected set of audit examples. For further information, there are now many written and consultative resources related to audits and auditing. Most professional and trade journals have articles on the subject, and most professional and trade organizations have developed special materials on auditing. The section in the Appendix entitled "Resources for SP" contains the names of organizations that can provide information on audits and auditing. The journals and magazines listed in the Appendix also are sources of information.

Every kind of business has its own special requirements for auditing. People in the real estate business and anyone else involved in the transfer of land and facilities, for example, use terms such as "site auditing," "environmental assessment," and "risk assessment." The purpose of these audits is to determine the degree to which potential problems and liabilities exist in the acquisition of land and facilities.

To provide some idea of the different kinds of audits, we can use the mining industry as an example. In a recent article about auditing in a major trade journal [11], the following kinds of audits were identified:

- Permit-performance audits, including a review of environmental quality-assurance plans, environmental permits, and restrictions governing operations.
- Regulatory-requirement audits to assess facility operations that are governed by local, state, and federal regulations.
- Environmental-management-practices audits to examine management structures, procedures, and policies used to implement compliance and communicate environmental awareness to the work force.
- Technical processes/practices audits to determine whether design or process modifications will accomplish environmental objectives to reduce emissions, waste, etc.
- Risk-management audits that will determine the probability of downstream problems, accidents, and liability.
- Special-purpose audits to respond to special needs or requirements such as determination of insurance liability or an EPA consent decree.
- Site assessments to examine past and current environmental hazards resulting from buried wastes, seepage from underground tanks, and toxic discharges into water systems.

It is beyond the scope of this book to cover all the possible types of audits. The audit is, however, a major tool for SP—a tool that can be designed to help an organization to go beyond compliance.

Audits in the U.S. first grew out of the requirement to show compliance with EPA and other regulations. After such catastrophes as Allied-Signal's dumping of the pesticide Kepone in Virginia's James River and Union Carbide's experience in Bhopal, India, auditing took on a wider purpose: to identify potential problems and risks and to correct processes and improve facilities to prevent such accidents. Auditing now is more than a compliance tool or a risk-avoidance tool. If auditing programs and audits are designed to help the organization to move toward SP, they will not only identify issues related to compliance and risk, they will identify opportunities to improve total quality and the organization's competitive position.

The Auditing Process

Audits have the most utility if they are embedded in an auditing program. They must reflect the organization's SP policy and the specific objectives of the auditing program. Useful audits are developed on the assumptions that management commitment exists and that the other steps described for creating an auditing program have been taken.

Audits may be conducted by people outside the organization; people who belong to a facility, system, or function; or a combination of the two. It is highly desirable that auditing become a routine way of doing business.

If audits always are conducted by outside people, then auditing becomes nothing more than the traditional quality-control system of "inspecting quality in." The goal of auditing is to make the people most closely involved in a function or system responsible for its environmental performance and the continuous improvement of that performance. It follows, therefore, that the people affected most directly by an audit should be directly involved in that audit. Their degree of involvement will vary according to the maturity of the organization's involvement in SP and its experience in conducting audits.

The steps for organizing and conducting an audit are

1. Defining the scope of the audit.
2. Establishing the audit team.
3. Preparing for the audit.
4. Conducting the audit.
5. Reporting findings and recommendations.

Defining the Scope of the Audit

The first step is to establish the purposes and parameters of the audit by setting its objectives and by identifying the specific organizational system, function, or site to be audited. The objectives of any specific audit should be clearly connected to the questions that have been identified as priorities in the overall auditing program.

Objectives are set so that the data-collection process will be clearly focused on determining *current performance* and *opportunities for change and improvement*.

The scope of an audit should permit the audit to be conducted efficiently and effectively. One way to limit the utility of an audit is to give it overblown objectives and embed it in a bureaucratic management system.

Establishing the Audit Team

Teams can be given objectives or they can be involved in the process of setting objectives. The composition of audit teams must ensure that the necessary competencies for conducting the audit are available and that the people affected by the audit are represented. Depending on the scope of the audit, the kinds of expertise required for a manufacturing organization may include the following:

- Chemical engineering
- Geology
- Civil engineering
- Environmental engineering
- System specialists

In addition to the above, the team might include people who have special or historical knowledge about a system, facility, or function.

Preparing for the Audit

The third phase in auditing is for the team to prepare to conduct the audit. The specific preparation required depends on the scope of the audit. The following are preparations required of most audit teams:

- Review scope of the audit.
- Set schedule and assign responsibilities.
- Review auditing procedures.
- Review data-collection process and documentation.
- Review reporting requirements.
- Review appropriate records and documents that describe the past and current performance of the system, function, facility or site being audited.
- Review previous audits and reports.
- Review all relevant regulatory requirements.
- Review plans to expand the system, function, or facility.
- Review known violations.
- Review environmental-management system and personnel responsibilities.

It is not possible to define all the specific requirements to prepare a team to conduct an audit. An audit of a paper mill will require information specific to that site, review of topographic maps, and review of the manufacturing process. An audit of a laboratory might require information about what chemicals are used and how radioactive materials are handled. Whatever the specific tasks needed to prepare for an audit happen to be, the general purposes are to take every action required to ensure the efficient and effective conduct of the audit.

One special task of an audit team is to ensure that there is a written protocol or manual for conducting the audit. The manual should include the scope of the audit, checklists, data-collection procedures, and reporting requirements. The manual is used not only by the audit team but also by the managers who are responsible for preparing their sites, functions, or systems for the audit.

Conducting the Audit

Audits that most directly support SP will involve directly the people who are affected by them. Audit teams will be composed of members of the work force and people with additional competencies from wherever they can be obtained. The goal of auditing is not to hide failure or even to find fault. The goal is continuous improvement of SP.

The conduct of an audit consists of the following steps:

1. Preaudit meetings to ensure that the team conducting the audit and all other people affected by the audit know what is going on.
2. Briefings and walk-throughs to familiarize outside members of the audit team with the audit target (site, production process, etc.).
3. Data collection.

The first two steps are self-explanatory. The third step requires some explanation.

Data can be collected in various ways. Some of the more common practices are

- Interviews using checklists contained in the audit manual.
- Measurement of key environmental variables (e.g., noise levels, emissions, deviations from standards, quality of discharge water, energy use).
- Visual inspection of facilities and practices (e.g., storage tanks, reclamation processes, waste disposal).
- Review of relevant documentation.

Reporting Findings and Recommendations

In discussing the design of an auditing program, I proposed that management should ensure that the program include a reporting process that

- Provides information that requires action.
- Stimulates and supports the exchange of information collected.
- Stimulates and encourages the involvement of the work force in improving the organization's SP.
- Stimulates and encourages the involvement of all other organizational stakeholders in improving the organization's level of SP.

The report from any one audit should meet at least the first three criteria. How all stakeholders should be involved in the results of the auditing program is not a decision that will be made relative to each audit.

To ensure that the audit requires action, the audit report should accomplish at least the following:

- Relate to the whole auditing program and the questions the program exists to answer.
- Describe findings related to the specific objectives of the audit.
- Identify areas of noncompliance with regulatory requirements.
- Identify areas of potential risk and future liability.
- Identify improvements that have been achieved since the previous audit.
- Make recommendations for new improvements and propose plans to achieve these improvements.

To ensure that the audit fully involves the people most affected by the audit, briefings on the results of the audit should be conducted for the people who can improve areas of performance identified in the audit. Briefings and other public forums also should be used to reward work teams that have contributed to performance improvements identified in the audit.

Auditing programs and auditing can be powerful tools to support SP. Auditing is a major tool for developing the performance baselines that organizations need in order to track their environmental performance and identify opportunities for improvement. If conducted according to the Principles of Sustainable Performance, audits can create greater involvement of the whole work force in the continuous improvement of the organization's performance—as well as the involvement of all other stakeholders.

Because audits vary so widely according to the nature of the organization's business and the specific variables being audited, it is not possible to describe of all the kinds of audits that could be used. However, examples of audit checklists are included in the Appendix to provide information about what the contents of specific audits might be.

The SP tools discussed thus far are the Sustainable-Performance Assessment and auditing. Two additional tools remain to be described: benchmarking and life-cycle analysis.

Benchmarking

Benchmarking has gained in popularity through the current emphasis on TQM. Benchmarking is a surveyor's term. A benchmark is a known height from which other elevations can be determined by comparison. Benchmarking, as used in TQM, is a tool for determining the best example of performance of a function or process and then comparing oneself to that benchmark or standard.

The process of developing and using benchmarks is easier to describe than it is to follow. Benchmarking, like all other total quality initiatives, assumes that management is fully committed to the process. Common to all benchmarking processes are steps such as the following.

1. Decide what to benchmark. Successful benchmarking has been reported for administrative processes, accounting processes, production processes, design sequences, human resource functions, and management practices. At a recent conference sponsored by the Global Environmental Management Initiative, participants were asked to identify areas related to environmental performance for which they would like to have the names of benchmark companies. It is apparent that, when SP is understood as the next evolution in total quality performance, benchmarking gains ever wider use for assessing environmental performance.

Benchmarks might be the best practices related to the following:

- Environmental management
- Environmental auditing
- Hazardous-waste management
- Emission reduction
- Packaging
- Reduction of the use of harmful chemicals
- Stakeholder involvement
- Control of fugitive emissions
- Reuse of secondary products and waste

2. Develop baselines and sufficient documentation of the performance to be benchmarked so that a comparison can be made with a benchmark. In the area of SP, environmental audits become a major source for such documentation.

3. Find the benchmark. Discovering benchmarks can be a complicated and time-consuming process. One new resource that gives promise of facilitating the search for benchmarks is the International Benchmarking Clearinghouse, which is part of the American Productivity and Quality Center in Houston, Texas. The Clearinghouse will assist an organization in a variety of ways to conduct a benchmarking process, including doing the research to find organizations that represent the best practices in the area of interest.

Additional sources of information for benchmarks include the following:

- Information from customers and suppliers
- Trade publications—including editors and publishers
- Trade and professional societies
- Quality award winners (e.g., Honor Role Awards given by the National Environmental Association,

the National Recycling Coalition awards, the National Wildlife Federation's Corporate Conservation Council Environmental Achievement Award, the U.S. Environmental Agency Administrator's Award, the President's Environment and Conservation Challenge Awards, and the World Environment Center's Gold Medal Award)
- Annual reports and other company publications
- Books and newspapers
- Consultants

Searches of the literature now can be greatly facilitated by the use of various databases such as *Trade and Industry ASAP, Dow Jones News, Business and Industry News,* and *Annual Reports Abstract.*

4. Compare the parent organization's performance with that of the benchmark organizations. This step may uncover the fact that the measurement data are not compatible, and the parent organization must retrieve new data or develop new measurement procedures.

5. Develop improvement objectives and improvement projects to achieve these objectives. Improvement objectives usually are most useful if they represent incremental change over a period of time.

6. Review and revise the benchmarking process. This may include developing new benchmarking objectives, new benchmarks, or new improvement objectives.

One common benchmarking process that has been developed is for one or more organizations to work together to find the best practice to serve as a benchmark. In 1991, Intel and AT&T teamed up to find the benchmark for corporate pollution programs. They identified the key subsystems and elements in a corporate pollution program and then identified benchmarks for each of these elements. The elements included paper recycling, Superfund management, auditing process, and so on. The best organizations that they identified as benchmarks included 3M, Dow, DuPont, H.B. Fuller, and Xerox. Benchmarks became the bases for revising the pollution-control systems in AT&T and Intel [8].

More and more, benchmarking is becoming a part of the standard practices of organizations that are committed to the continuous improvement of total quality—including environmental quality. As such, benchmarking is a tool that will be of use in all organizations that commit to SP.

Life-Cycle Analysis

The environmental impact of a product does not begin at the postconsumer stage—after it has been used. The total environmental impact of a product begins at the point at which materials are extracted from their sources (e.g., the ground) and ends with the final output into the environment in the form of pollution, waste, and emission. In between the point of extracting the raw materials and the final disposal of all wastes, effects on the environment occur through all stages of processing, manufacturing, packaging, transportation, and consumption.

The Systems Model for Sustainable Performance (Figure 4-1) provides a framework for understanding the underlying methodology of life-cycle analysis. The production of any product, whether through manufacturing or agriculture, can be considered a system with multiple inputs from the environment and multiple outputs to the environment. Life-cycle analysis (LCA) is a process of analyzing all these inputs and outputs to determine the total environmental impact of the production and use of a product. For example, the amount of energy used along the whole life cycle of a product is one primary variable that is measured and analyzed.

LCA has a number of advantages over the traditional "end-of-pipe" assessment of environmental impact and the large-scale descriptions of environmental issues such as greenhouse effect, ozone depletion, deforestation, and desertification. LCA employs a set of inventories that quantifies use of energy, use of resources, and environmental releases into the air, water, and land. LCA produces quantifiable information that can

- Provide a comprehensive picture of the environmental impact of any product (as opposed to the simplistic pictures produced by false alternatives offered between paper and plastic, coal and oil, etc.).
- Assist in identifying reasonable tradeoffs in product and process selection.
- Help to develop baselines against which future decisions can be made about modifying a product.

Phases

If we analyze the production of any product within the framework of the Systems Model for Sustainable Performance, we can develop a set of stages or phases that an application of LCA will include. These phases are

1. Analysis of the input of raw materials into a production process.
2. Analysis of the processing or preparation of raw materials for use in the process.
3. Analysis of the production process.
4. Analysis of the packaging process.

5. Analysis of the transportation and distribution process.
6. Analysis of reclamation of waste and secondary products.
7. Analysis of waste management.

Figure 7-1 shows how each of these phases is related and how each phase represents a step in analyzing total input from the environment and total output to the environment. The inputs analyzed include materials, water, air, and energy. The outputs include atmospheric emissions, waterborne wastes, solid wastes, hazardous wastes, and reclaimed products.

Limitations

The LCA is a tool and does not provide all the information required to manage for SP. LCAs do not provide comparative information among different processes that differ widely in the kinds of materials they use and in the kinds of wastes or emissions they produce. Also, LCAs cannot compare the relative environmental merits of inputs to outputs. They cannot tell us if it is better to use a certain kind of raw material and produce a certain kind of waste or to select a different kind of raw material in order to produce a different kind of waste.

Example

The Council on Plastics and Packaging in the Environment reports a typical application of LCA. High-density polyethylene (HDPE) is used in the typical 1,000 liter bottle for a liquid fabric-conditioner product. LCA was used to compare the relative merits of several optional packaging modifications being considered. These were incorporating 25 percent recycled HDPE into the virgin HDPE bottle;

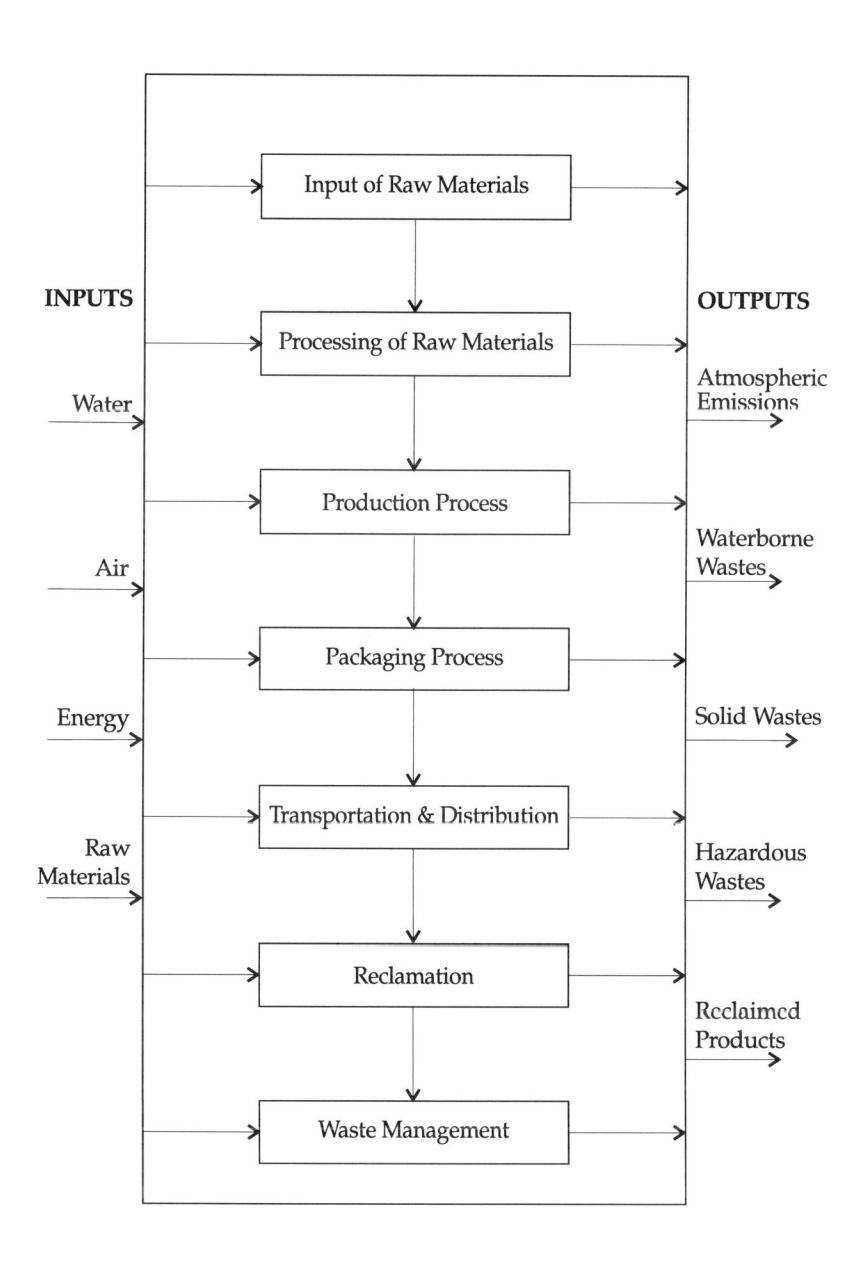

Figure 7-1. Phases in the Life-Cycle-Analysis Process

using a smaller bottle with concentrated fabric conditioner; and using a paper carton with concentrated refill. Contrary to what common sense might suggest, the LCA demonstrated that the concentration of the product afforded greater reductions in energy use and emissions than did the use of 25 percent recycled HDPE.

Summary

In this chapter, I have described the general strategies for SP as well as selected tools. There are, of course, several other strategies and tools that are coming into more common use, which I have not considered of sufficient general use to include in this book. Examples of these are closed-loop processing and design-for-environment.

I have discussed the strategies and tools as sources of information that are required to manage SP. These information sources should be used by leaders to develop the competencies required for managing SP.

In the next and final chapter, I will review the Sustainable-Performance Management Model and suggest how the reader can "put it all together" and integrate what I have presented into a practical understanding for managing SP.

References

1. Barriers Facing the Achievement of Ecologically Sustainable Industrial Performance. (1991). *Proceedings of the Conference on Ecologically Sustainable Industrial Development*. Vienna, Austria: United Nations Industrial Development Organization.
2. *Business and Environment*. (October 25, 1991). p. 4.
3. *Business Month Magazine*. (May, 1990). pp. 50-51.
4. Callenbach, E., Capra, F., & Manburg, S. (1990). *Eco-auditing and Ecologically Conscious Management*. Berkeley, CA: Elmwood Institute.
5. *The Economist*. (September 8, 1990). p. 10.
6. Frosch, R., & Gallopoulos, N. (September, 1989). Strategies for Manufacturing. *Scientific American*, pp. 94-102.
7. International Chamber of Commerce. (1991). *ICC Guide to Effective Environmental Auditing*. Paris: ICC Publishing S.A.
8. Klafter, B. (1992). Case Study: AT&T and Intel Pollution Prevention Benchmarking. *Proceedings of Corporate Quality/Environmental Management II*. Washington, D.C.: Global Environmental Management Initiative.
9. Newman, S. (June, 1990). Return to Vendor: Aluminum Recycling. *Management Review*, p. 21.
10. Office of Technology Assessment (1986). *Serious Reduction of Hazardous Waste: For Pollution Prevention and Industrial Efficiency*. Washington, D.C.: Author.
11. Philbrook, J. (February 1991). Environmental Audits: Determining the Need at Mining Facilities. *Mining Engineering (43)*, pp. 207-209.

12. Schecter, R., & Hunt, G. (1989). *Case Summaries of Waste Reduction by Industries in the Southeast.* Raleigh, NC: Waste Reduction Resource Center for the Southeast.
13. Springer, J. (1992). *Pollution Case Studies Compendium.* Washington, D.C.: Environmental Protection Agency.
14. Stratton, B. (April, 1991). Going Beyond Pollution Control. *Quality Progress,* pp. 18-20.

8

Putting It All Together

This book is about the new rules under which organizations will be required to function in the age of the environment. It also is about the concrete steps that organizations must take to respond to these new rules. In the foregoing pages, I have introduced the idea of sustainable performance as one that can serve as an organizing concept and tool for planning and shaping an organization's response to the environmental challenge.

In drawing this book to a close, I would like to do the following:

1. Emphasize the new rules under which organizations must learn to operate.
2. Emphasize, in a summary way, what leaders can do to (a) help their organizations learn to play by the new green rules, and (b) move their organizations toward sustainable performance.

The New Rules Are Green

Leaders of organizations have been called on to respond to various "new rules" over the past fifty years. The work force has changed in its demographics and often in its values. Traditional control models for managing people have proven inadequate. Competition has shifted from national to international markets. Quality no longer means just performance against preset standards. It means performance that produces totally satisfied customers.

The new rule that includes all other new rules is that the continuous improvement of every aspect of business is now a way of life. Organizations that have not learned to live by this rule have suffered in the marketplace. Published anecdotes about organizations that, a short time ago, were considered "excellent" are now objects of curiosity more than they are sources of useful information.

The newest rules are green rules. The continued survival of organizations now depends on their learning to play by the green rules. Although I have not specifically listed them as rules, I have covered these rules in the process of describing the meaning of sustainable performance and what organizations must do to make their performance sustainable. If organizations want to stay competitive and alive, they must learn to play by these green rules:

1. The "greening" of business is inevitable. A host of pressures are mounting that are forcing organizations to improve their environmental performance. Businesses will either be green or they will be dead.
2. The environment now must be considered an organization's most important supplier and most valued customer. Quality in performance cannot be achieved except by 100 percent customer satisfaction, and this means that the environment is 100 percent satisfied.
3. The sooner that organizations begin to view the environmental challenge as a competitive opportunity, the more likely it is that they will stay alive and profit.

The pressure that organizations are now feeling to demonstrate their capacity to conduct their business while actively reducing environmental problems and supporting environmental causes, though currently great, can be expected to intensify rapidly within the immediate future. With ever increasing frequency, environmental goals and strategies will appear in strategic plans and in the annual reports of corporations to their stockholders. Organizations that have not already done so will have highly positioned officers and managers with environmental titles and responsibilities. Performance-appraisal processes will be revised to show that people are held responsible for the quality of their performance toward the environment. Within a very few years, the organization that does not have a TQEM program (by whatever name it may be called) will be the exception in the same way that the organization that does not have a TQM program today is the exception.

The demand to respond responsibly toward the environment is not one that will pass with time. Although the earth's ecosystems have proven to be extremely resilient thus far, it is apparent that we humans have put such pressure on these systems that their capacity to repair themselves has now been overreached.

The question that all organizations face is not whether they will go green. The only choice that organizations have at the moment is *when*. For a time, it will be possible to "appear" green, but the hard facts of competition and survival are quickly driving organizations to embrace the environment as their new and most important customer.

Organizations that already have discovered both the imperative and the value of going green have introduced extensive initiatives to change the ways in which they do business. 3M Corporation entered the environmental competition early with its 3Ps program (Pollution Prevention Pays). Dow Chemical has introduced WRAP (Waste Reduction Always Pays). Chevron has SMART (Save Money and Reduce Toxics). General Electric has its POWER program (Pollution, Waste, and Emissions Reduction).

It has not been just the largest organizations or those in the manufacturing and chemical industries that have learned to play by the new rules. All kinds of organizations are learning—large and small, national and international.

The German Environmental Management Association (B.A.U.M.) is composed primarily of small businesses. The head of B.A.U.M. recently noted at an international conference that in Germany

> There is an increasing number of companies that are gaining market success by changing over to environmentally sounder products and reducing cost of energy, water and raw materials savings programmes [8].

STORA is the largest forest-products company in Europe. From 1975 to the present, its Swedish pulp and paper mills decreased their emissions of sulphur dioxide by more than 85 percent. At the same time, the production of pulp increased by 25 percent [1].

Imperial Chemical Industries (ICC) is the fifth-largest chemical manufacturing company in the world. One of its important products is paints for the automotive industry. Toxic emissions from the use of paints account for a large amount of the total discharge of volatile organic compounds into the air. ICC has responded to the needs of the automobile industry to reduce its emissions by developing a group of waterborne or water-based paints. These paints are finding growing acceptance in the automobile industry while serving the needs of the environment and ensuring ICC's future market success [3].

One suggestion of how organizations are learning to play by the new green rules is the information that they are putting in their annual reports. The following are a few examples.

In its annual report for 1992, Kansas Power and Light Company maintains that its "commitment is deeper than just meeting legal requirements; environmentalism is part of the corporate culture." Employees from across the corporation serve on a task force that both proposes and oversees environmental projects that range from recycling to preservation of wetlands.

Borden, Inc., adopted *Principles of Environmental Responsibility* and published them in 1991. The company also created the position of Vice President, Health and Environment, to monitor more closely its efforts to protect the environment. Borden voluntarily joined the EPA's Air Toxics Reduction Program and committed to reducing its airborne emissions by 33 percent by year-end 1992 and by 50 percent by year-end 1994. At its North American forest-products adhesive plants, Borden is aiming at zero discharge of pollutants. Five plants already have achieved this goal. Borden also has reduced the amount of packaging material used in its dairy, pasta, and food-service products. Most paperboard used in its pasta boxes is 100 percent recycled.

Orange and Rockland Utilities, Inc., of Pearl River, New York, maintains in its 1992 annual report that a key component of its corporate strategy is to "Operate in an environmentally sound manner, meeting all standards, engaging in environmental activities consistent with responsible corporate citizenship and recognizing environmental benefits in evaluating future resources."

Businesses quite different from those in the manufacturing, chemical, retail, and utilities industries are learning to play by the green rules. Innkeeping, the golfing industry, and even the U.S. Postal Service are going green.

Major hoteliers are vying for the "green rights" in their industry. The International Hotel Association has its own environmental award. The award is based largely on energy saving. In 1991, it was won by Ramada International Hotels and Resorts for its "Hotels of the New Wave" program, which it operates in 110 hotels in thirty-six countries. Ramada also has unveiled a campaign to raise money for The Nature Conservancy. Together with American Express, the group will give one dollar to The Nature Conservancy each time a guest of Ramada pays with an American Express Card.

Marriott's southern California hotels have announced that they will invest in more efficient lighting and save 20 percent on lighting energy.

Loews Hotels has launched its "Good Neighbor" program. The chain will recycle paper and install energy-saving equipment.

Westin Hotels and Resorts has a number of environmental initiatives underway, including water and energy conservation, waste recycling, and donating food and other surplus supplies to charity—rather than throwing them away as more waste for landfills [6].

In February, 1992, the Golf Course Superintendents Association of America (GCSAA) devoted an entire issue of its *Golf Course Management* magazine to an environmental update [4]. The GCSAA now offers an Environmental Management Program and actively supports a variety of initiatives to create golf courses that are friendly toward the environment.

The U.S. Postal Service now has an environmental policy that pledges it to

...use...nonpolluting technologies and waste minimization in the development of equipment, products, and operations....Promote the sustainable use of natural resources....Include environmental considerations among the criteria by which new projects, products, processes, and purchases are evaluated.

The U.S. Postal Service has launched the nation's largest recycling program involving nearly 730,000 employees and 40,000 postal facilities. It now operates the nation's largest alternative-fuel delivery fleet, using compressed natural gas. The Postal Service also publishes pamphlets and booklets that provide the public with ways to be more environmentally responsible, e.g., "Mail That Mother Earth Can Love" [7].

Playing by the Green Rules

Frank Popoff, president and chief executive officer of Dow Chemical, in a 1990 speech made to the International Chamber of Commerce, suggested that the chemical industry's response to environmental concerns had gone through three stages: "Denial, Data, and Dialogue" (the three Ds). The third D, dialogue, characterizes the stage that Popoff believes most organizations are entering. This is the stage of communication and ongoing efforts to become part of the solution to the world's environmental problems. In his speech, however, Popoff called for his colleagues to move even further to a fourth D: "Doing the right thing." This stage, according to Popoff, would include the following:

- Looking for opportunities to enhance sustainable development.
- Prioritizing and dedicating capital and resources to be spent on environmental practices, with emphasis

on continuous improvement in environmental performance.
- Seeking ways to encourage industry associations to develop self-help programs similar to the *Responsible Care Program* of the chemical industry.
- Becoming more open and responsive to the public and forming cooperative partnerships.
- Giving employees and all constituencies opportunities to be practicing environmentalists [5].

I have maintained in this book that "doing the right thing" will prove to mean that "doing the right thing for the environment also is doing the right thing for the organization." I have expanded the meaning of "doing the right thing" to mean working toward sustainable performance and have proposed models and tools that can be used by leaders to move their organizations toward sustainable performance.

What Leaders Can Do

I have written a great deal in this book about what leaders can do to help their organizations move toward sustainable performance. Underlying everything that I have written about leadership is the assumption that leaders have three responsibilities: They must understand, plan, and act.

Understand

First, leaders must understand. To understand the great complexity associated with environmental issues and environmental performance is no small task. The idea of sustainable performance can assist in such understanding.

In this book, I have provided considerable information about the meaning of sustainable performance. I have proposed that it incorporates the characteristics of equity and stewardship, that it recognizes necessary limits to growth and the use of resources, and that it requires communal action and systemic thinking (Chapter 4).

I have proposed that to reach sustainable performance, an organization must, itself, become an ecological system that interacts with the natural environment's ecological systems in a benign way (Chapter 4).

I also have suggested that sustainable performance represents a radical shift in organizational values and have translated these values into a set of principles (Chapter 5).

Sustainable performance is a way of redefining the meaning of quality. It is the next step in the evolution of the meaning of quality performance. When leaders grasp this meaning, they will be ready to help their organizations plan their sustainable-performance initiatives.

Plan

The second thing that leaders can do is plan. The Sustainable-Performance Management Model (Chapter 1) is a tool for planning an organization's sustainable-performance initiative. The model identifies the essential milestones that must be achieved in a process that leads to sustainable performance. The most important milestone is the first, developing a sustainable-performance policy statement. The policy statement establishes ground rules and expectations for the organization by

- Clarifying for the organization exactly where the organization stands regarding the environment and its business interests.
- Integrating concerns for the environment with the strategic business interests of the organization.

- Focusing the organization's attention on those few interests that are crucial to its success.

Act

The third responsibility of management is to act. Acting to achieve sustainable performance does not mean acting alone. Leadership for sustainable performance, like leadership for total quality, is not a job for isolates. There are three attributes of leadership for sustainable performance that should be associated with every action leaders take.

1. Demonstrated commitment to sustainable performance.
2. Determination to involve the work force.
3. Determination to involve the organization's other stakeholders.

Action requires a knowledge of the various strategies and tools that can support the achievement of sustainable performance. In Chapter 7, I indicate that leaders must be knowledgeable of at least the following general strategies for sustainable performance:

1. Practicing conservation and paying attention to every detail associated with a work process, e.g., using only the necessary amounts of materials, turning off the water, turning off the lights, and keeping machinery and vehicles in top running condition.
2. Modifying or replacing existing processes, products, and services to make these environmentally friendly, e.g., changing to more energy-efficient machinery, reducing packaging materials, eliminating toxic chemicals and emissions.

3. Reclaiming by recycling and reusing waste and secondary products such as chemicals, paper, plastic, metal, and water.
4. Reducing the use of materials, e.g., reducing the amount of packaging or packing used, reducing the size of reports and invoices, reducing the amount of a material used in a process, and reducing energy use.

In addition to these strategies, there are specific technologies or tools that leaders should understand (Chapter 7).

- The Sustainable-Performance Assessment
- Auditing
- Benchmarking
- Life-Cycle Analysis

These are the requirements of leadership for sustainable performance: understand, plan, and act. The greatest of these (by far) is to understand.

Conclusion

The source for economic strength, profitability, and creative technology is business. Business is the repository of the research, technology, capital, and managerial competence that is required to resolve the environmental crisis.

Business also commands the fiscal and material resources to resolve the environmental crisis. This fiscal power is most obvious in the case of transnational or international corporations. General Motors (U.S.), for example, has approximately the same annual income as Australia; Siemens (Germany) has the same income as Algeria; and Peugeot (France) has the same income as Peru [2].

Our hope for saving the environment is to do everything we can to show business and industry that the current challenge of the environment should be accepted wholeheartedly as an opportunity to reach new plateaus of profitable performance.

If the question is how to save our natural environment, then we already know the answer. We will save the environment by the way we run our companies and conduct our business.

Our problem with saving our environment is not that we do not know the answer, it is that we have yet to accept the obvious answer and act on it. *Our most promising opportunity to save the environment is to enroll the full spectrum of business and industry in working toward sustainable performance.*

References

1. Berggren, B. (1990). An Industrialist's View on Industry's Environmental Responsibilities. In J. Willums (Ed.), *The Greening of Enterprise.* Paris: International Chamber of Commerce.
2. Ekins, P. (1992). *Green Economics.* New York: Anchor Books.
3. Elkington, J., & Burke, T. (1989). *The Green Capitalist.* London: Victor Gollancz, Ltd.
4. Environmental Update. (February, 1992). *Golf Course Management.*
5. Kirkpatrick, D. (February 12, 1990). Environmentalism. *Fortune*, pp. 44-52.
6. Maney, K. (January 20, 1992). Hotel Chains Get Green Light. *USA Today*, p. 6B.

7. U.S. Postal Service. (1991). *Mail That Mother Earth Can Love.* Washington, D.C.: Author.
8. Winter, G. (1990). The Challenge of Environmental Management. In J. Willums (Ed.), *The Greening of Enterprise.* Paris: International Chamber of Commerce.

APPENDIX

This Appendix has the following sections:

- **Section I: International Environmental Business Codes.** This section contains the *CERES Principles* and the *International Chamber of Commerce Business Charter*.
- **Section II: Resources for Sustainable Performance.** This section lists the addresses of various organizations that are resources for information, publications, and consultative services that are available to individuals and organizations. The list is international in scope.
- **Section III: Environmental Laws.** The laws described are the primary U.S. Federal laws. Also included are resources for finding information about the environmental laws of many other countries.
- **Section IV: Journals and Magazines.** Most business publications now carry, from time to time, articles about the environment. The journals and magazines listed in this section are limited to those that have a primary focus on the environment or that consistently include articles about business and the environment in their issues.
- **Section V: Environmental-Audit Checklists.** The kinds of audit checklists that are being used are now

extremely large and vary greatly in focus and content. The lists contained in this section serve only to introduce the reader to the nature of environmental-audit checklists.

- **Section VI: The Sustainable-Performance Assessment (SPA)**. The SPA measures the performance levels that organizations have reached in their process toward full sustainable performance. The SPA is a tool that organizations can use to develop a baseline from which they can track their progress and by which they can identify opportunities for improving performance.

SECTION I: INTERNATIONAL ENVIRONMENTAL BUSINESS CODES

CERES

The Coalition for Environmentally Responsible Economies (CERES) is a nonprofit, membership organization that is composed of leading social investors, environmental groups, religious organizations, public-pension trustees, and public-interest groups. The Coalition met first in early 1989 to focus on ways that investors could help to implement environmentally and financially sound investment policies.

First introduced as the *Valdez Principles* in 1989, the CERES Principles are a ten-point code (or environmental ethic) for corporations. The Principles were devised to encourage the development of positive programs designed to prevent environmental degradation, to assist organizations in setting policy, and to enable investors to make informed choices regarding environmental issues.

The CERES Principles*

Introduction

By adopting these Principles, we publicly affirm our belief that corporations have a responsibility for the environment, and must conduct all aspects of their business as responsible stewards of the environment by operating in a manner that protects the Earth. We believe that corporations must not compromise the ability of future generations to sustain themselves.

We will update our practices constantly in light of advances in technology and new understandings in health and environmental science. In collaboration with CERES, we will promote a dynamic process to ensure that the Principles are interpreted in a way that accommodates changing technologies and environmental realities. We intend to make consistent, measurable progress in implementing these Principles and to apply them to all aspects of our operations throughout the world.

Protection of the Biosphere

We will reduce and make continual progress toward eliminating the release of any substance that may cause environmental damage to the air, water, or the earth or its inhabitants. We will safeguard all habitats affected by our operations and will protect open spaces and wilderness, while preserving biodiversity.

Sustainable Use of Natural Resources

We will make sustainable use of renewable resources, such as water, soils and forests. We will conserve nonrenewable natural resources through efficient use and careful planning.

---

*Reprinted by permission of the Coalition for Environmentally Responsible Economies.

Reduction and Disposal of Wastes

We will reduce and where possible eliminate waste through source reduction and recycling. All waste will be handled and disposed of through safe and responsible methods.

Energy Conservation

We will conserve energy and improve the energy efficiency of our internal operations and of the goods and services we sell. We will make every effort to use environmentally safe and sustainable energy resources.

Risk Reduction

We will strive to minimize the environmental, health and safety risks to our employees and the communities in which we operate through safe technologies, facilities and operation procedures, and by being prepared for emergencies.

Safe Products and Services

We will reduce and where possible eliminate the use, manufacture or sale of products and services that cause environmental damage or health or safety hazards. We will inform our customers of the environmental impacts of our products or services and try to correct unsafe use.

Environmental Restoration

We will promptly and responsibly correct conditions we have caused that endanger health, safety or the environment. To the extent feasible, we will redress injuries we have caused to persons or damage we have caused to the environment and will restore the environment.

Informing the Public

We will inform in a timely manner everyone who may be affected by conditions caused by our company that might endanger health, safety or the environment. We will regularly seek advice and counsel through dialogue with persons in communities near our facilities. We will not take any action against employees for reporting dangerous incidents or conditions to management or to appropriate authorities.

Management Commitment

We will implement these Principles and sustain a process that ensures that the Board of Directors and Chief Executive Officer are fully informed about pertinent environmental issues and are fully responsible for environmental policy. In selecting our Board of Directors, we will consider demonstrated environmental commitment as a factor.

Audits and Reports

We will conduct an annual self-evaluation of our progress in implementing these Principles. We will support the timely creation of generally accepted environmental audit procedures. We will annually complete the CERES Report, which will be made available to the public.

International Chamber of Commerce

The International Chamber of Commerce formally launched *The Business Charter for Sustainable Development* in April, 1991, at the Second World Industry Conference on Environmental Management.

The Business Charter for Sustainable Development*

Introduction

Sustainable development involves meeting the needs of the present without compromising the ability of future generations to meet their own needs.

Economic growth provides the conditions in which protection of the environment can best be achieved, and environmental protection, in balance with other human goals, is necessary to achieve growth that is sustainable.

In turn, versatile, dynamic, responsive and profitable businesses are required as the driving force for sustainable economic development and for providing managerial, technical and financial resources to contribute to the resolution of environmental challenges. Market economies, characterized by entrepreneurial initiatives, are essential to achieving this.

Business thus shares the view that there should be a common goal, not conflict, between economic development and environmental protection, both now and for future generations.

Making market forces work in this way to protect and improve the quality of the environment—with the help of performance-based standards and judicious use of economic instruments in a harmonious regulatory framework—is one of the greatest challenges that the world faces in the next decade.

The 1987 report of the World Commission on Environment and Development, "Our Common Future," expresses the same challenge and calls on the cooperation of business in tackling it. To this end, business leaders have launched actions in their individual enterprises as well as through sectoral and cross-sectoral associations.

---

*Reprinted by permission of the International Chamber of Commerce.

In order that more businesses join this effort and that their environmental performance continues to improve, the International Chamber of Commerce hereby calls upon enterprises and their associations to use the following Principles as a basis for pursuing such improvement and to express publicly support for them.

Principles

1. **Corporate priority.** To recognize environmental management as among the highest corporate priorities, and as a key determinant to sustainable development; to establish policies, programmes and practices for conducting operations in an environmentally sound manner.

2. **Integrated management.** To integrate these policies, programmes and practices fully into each business as an essential element of management in all its functions.

3. **Process of improvement.** To continue to improve corporate policies, programmes and environmental performance, taking into account technical developments, scientific understanding, consumer needs and community expectations, with legal regulations as a starting point; and to apply the same environmental criteria internationally.

4. **Employee education.** To educate, train and motivate employees to conduct their activities in an environmentally responsible manner.

5. **Prior assessment.** To assess environmental impacts before starting a new activity or project and before decommissioning a facility or leaving a site.

6. **Products and services.** To develop and provide products or services that have no undue environmental impact and are safe in their intended use, that are efficient in their consumption of energy and natural resources, and that can be recycled, reused, or disposed of safely.

7. **Customer advice.** To advise, and where relevant educate, customers, distributors, and the public in the safe use, transportation, storage and disposal of products provided; and to apply similar considerations to the provision of services.
8. **Facilities and operations.** To develop, design and operate facilities and conduct activities taking into consideration the efficient use of energy and materials, the sustainable use of renewable resources, the minimisation of adverse environmental impact and waste generation, and the safe and responsible disposal of residual wastes.
9. **Research.** To conduct or support research on the environmental impacts of raw materials, products, processes, emissions and wastes associated with the enterprise and on the means of minimizing such adverse impacts.
10. **Precautionary approach.** To modify the manufacture, marketing or use of products or services or the conduct of activities, consistent with scientific and technical understanding, to prevent serious or irreversible environmental degradation.
11. **Contractors and suppliers.** To promote the adoption of these principles by contractors acting on behalf of the enterprise, encouraging and, where appropriate, requiring improvements in their practices to make them consistent with those of the enterprise; and to encourage the wider adoption of these principles by suppliers.
12. **Emergency preparedness.** To develop and maintain, where significant hazards exist, emergency preparedness plans in conjunction with the emergency services, relevant authorities and the local community, recognizing potential transboundary impacts.
13. **Transfer of technology.** To contribute to the transfer of environmentally sound technology and

management methods throughout the industrial and public sectors.

14. **Contributing to the common effort.** To contribute to the development of public policy and to business, governmental and intergovernmental programmes and educational initiatives that will enhance environmental awareness and protection.

15. **Openness to concerns.** To foster openness and dialogue with employees and the public, anticipating and responding to their concerns about the potential hazards and impacts of operations, products, wastes or services, including those of transboundary or global significance.

16. **Compliance and reporting.** To measure environmental performance; to conduct regular environmental audits and assessments of compliance with company requirements, legal requirements and these principles; and periodically to provide appropriate information to the Board of Directors, shareholders, employees, the authorities and the public.

Section II: Resources for Sustainable Performance

In order to assist readers from countries other than the United States, I have divided this section into two subsections. The first subsection lists organizations that are international in scope, with a number of resources from specific countries. The second subsection includes resources that are of primary relevance to business in the U.S. Given the fact that so many organizations are international and so much business is conducted on an international scale, the separation of these sources is somewhat arbitrary.

The following publications may be consulted for extensive lists of organizations that can serve as resources for sustainable performance:

- Lanier-Graham, S. (1991). *The Nature Directory*. New York: Walker and Company.
- Levine, M. (1991). *The Environmental Address Book*. New York: Perigree Books.
- Stein, E. (1992). *The Environmental Source Book*. New York: Lyons & Burford.

International Resources

- Business Council for Sustainable Development, World Trade Centre Building, 3rd Floor, 10 route de l'Aeroport, Geneva 15, Switzerland
- General Agreement on Trade and Tariffs (GATT), Rue de Lausanne 154, Geneva 1202, Switzerland
- International Chamber of Commerce, 38 Cours Albert 1er, F-75008, Paris, France
- International Environmental Bureau, Elveveien 25B, P.O. Box 301, N-1324 Lysaker, Norway
- International Environmental Information System, Department of the Environment, Department of Transport, Library Service, 2 Marsham Street, London SW1P 3EB, England
- International Network for Environmental Management, INEM, Sillemstrasse 36, W-2000 Hamburg 20, Germany
- Organization for Economic Cooperation and Development, 2 rue Andre' Pascal, Paris, France
- SustainAbility Limited, The People's Hall, 91-97 Freston Road, London W11 4BD, England

- UN Center for Transnational Corporations, Room DC-2-1202, United Nations, New York, NY 10017, U.S.A.
- UN Development Programme, United Nations, New York, NY 10017, U.S.A.
- UN Economic Commission for Europe, 8-14 Avenue de la Pax, 1211 Geneva, Switzerland
- UN Environment Programme, P.O. Box 30552, Nairobi, Kenya
- UN Environment Programme Industry & Environment Office, 39-43 Quai Andre Citroen, 75739 Paris Cedex 15, Paris, France
- UN Industrial Development Organization, P.O. Box 300, Vienna International Centre, 1400 Vienna, Austria
- World Environment Center, 419 Park Avenue South, Suite 1403, New York, NY 10016, U.S.A.
- World Resources Institute, 1709 New York Avenue, N.W., Washington, DC 20026, U.S.A.
- Worldwatch Institute, 1776 Massachusetts Avenue, N.W., Washington, DC 20036, U.S.A.

U.S. Resources

- Agency for International Development, 320 21st Street, N.W., Washington, DC 20523
- Agricultural Resources Center, 115 W. Main Street, Carrboro, NC 27510
- Alliance to Save Energy, 1725 K Street, N.W., Washington, DC 20026

- American Council for an Energy-Efficient Economy, 1001 Connecticut Avenue, N.W., Washington, DC 20036
- Clean Sites, Inc., 1199 North Fairfax Street, Suite 400, Alexandria, VA 22314
- Coalition for Environmentally Responsible Economies, 771 Atlantic Avenue, 5th Floor, Boston, MA 02111
- Elmwood Institute, 2642 College Avenue, Berkeley, CA 94705
- Environmental Defense Fund, 257 Park Avenue South, New York, NY 10010
- Environmental Law Institute, 1616 P Street, N.W., Washington, DC 20036
- Environmental Protection Agency, 401 M Street, S.W., Washington, DC 20460
- Global Environmental Management Initiative, 1828 L Street, N.W., Suite 711, Washington, DC 20036
- Green Cross Certification Company, 1611 Telegraph Avenue, Suite 1111, Oakland, CA 94612
- Green Seal, P.O. Box 1694, Palo Alto, CA 94320
- Institute for Alternative Agriculture, Inc., 9200 Edmonston Road, Suite 117, Greenbelt, MD 20770
- International Alliance for Sustainable Agriculture, University of Minnesota, 1701 University Avenue, S.E., Newman Center, Room 202, Minneapolis, MN 55414
- Investor Responsibility Research Center, Suite 600, 1755 Massachusetts Avenue, N.W., Washington, DC 20036
- Management Institute for Environment and Business, 1220 16th Street, N.W., Washington, DC 20036

- National Solid Wastes Management Association, 1730 Rhode Island Avenue, N.W., Suite 1000, Washington, DC 20036
- National Wildlife Federation Corporate Conservation Council, 1400 16th Street, N.W., Washington, DC 20036
- Renew America, 1400 16th Street, N.W., Suite 710, Washington, DC 20036
- Resources for the Future, 1616 P Street, N.W., Washington, DC 20036
- Responsible Care Program, 2501 M Street, N.W., Washington, DC 20037
- Rocky Mountain Institute, 1739 Snowmass Creek Road, Old Snowmass, CO 81654
- Soil and Water Conservation Society, 7515 Northeast Ankeny Road, Ankeny, IA 50021
- Tufts University Center for Environmental Management, Tufts University, Curtis Hall, 474 Boston Avenue, Medford, MA 02155
- Union of Concerned Scientists, 26 Church Street, Cambridge, MA 02238

SECTION III: ENVIRONMENTAL LAWS

Summarized in this section are the principal U.S. Federal laws that impact on the way organizations do business. At the end of this section are resources for finding out more about the environmental laws of other countries.

U.S. Laws

Clean Air Act

An extraordinarily comprehensive law that is intended to control and prevent air pollution, the act regulates the emission into the atmosphere of any substance that affects the quality of the air. Regulations developed with the act govern pollutants such as nitrous oxides, sulfur dioxide, and carbon dioxide.

Community Response and Right-to-Know Act

This act mandates that all facilities that produce, transport, store, use, or release hazardous substances (as identified by EPA) provide full information to all local and state enforcement and safety authorities. The act further requires that such facilities maintain emergency-action plans to respond to unpermitted releases of hazardous materials.

Comprehensive Environmental Response, Compensation, and Liability Act (CERCLA)

Popularly referred to as the "Superfund," this act regulates the cleanup of disposal sites that were created before the enactment of the Resource Conservation and Recovery Act. It applies to incidents involving a release or substantial threat of a release of hazardous substances, pollutants, or contaminants into the environment. This act establishes potential liability for any person or organization having any present or historical responsibility for the creation of a site posing an actual or potential environmental hazard. Under CERCLA, individuals and organizations may be prosecuted, fined, or taxed to fund cleanup activities.

CERCLA was reauthorized under the "Superfund Amendment and Reauthorizations Act" (SARA). This act included a right-to-know provision and enlarged the regulatory requirements of CERCLA.

Federal Hazardous Substances Act (FHSA)

This act is administered by the Consumer Product Safety Commission. The commission regulates all hazards to health and safety associated with consumer products and has the authority to recall any product deemed hazardous.

Federal Water Pollution Control Act (FWPCA)

Popularly referred to as the "Clean Water Act" (CWA), this act regulates all discharges into surface waters. This act drives an extensive regulatory program that limits the discharge of pollutants directly or indirectly into surface waters. It has direct impact on the performance and construction of public sewer and storm-sewer systems.

Federal Insecticide, Fungicide, and Rodenticide Act (FIFRA)

This law controls the distribution, sale, and use of pesticides, fungicides, and rodenticides. It requires that businesses involved in the distribution, sale, or use of these materials register their use with the EPA. The law regulates the use of these substances as well the applicators used with them.

Hazardous Materials Transportation Act (HMTA)

This act regulates the packaging, marketing, labeling, and placarding of shipments of hazardous materials. Hazards include flammability, toxicity, and radiation. The act requires that all materials be properly classified prior to shipment, that proper notification be given to cognizant authorities prior to shipment, and that full information about hazardous materials accompany shipments.

Occupational Safety and Health Act (OSHA)

This law controls health and safety in the workplace. It requires businesses to comply with a wide range of standards that are designed to keep the workplace free from hazards to the health and safety of workers. The objective of the law is to ensure that workers do not suffer impairment in health or functional capacity.

Resource Conservation and Recovery Act (RCA)

This act regulates the generation, treatment, and disposal of solid and hazardous wastes from "cradle to grave." The act was amended by the "Hazardous and Solid Waste Amendments Act" (HSWA). This act brought thousands of small-quantity generators of hazardous waste—who previously had been exempt—under the law.

Both solid and hazardous wastes are regulated by RCA. The goal of the act is to ensure environmentally sound practices in waste management, recycling, and resource conservation.

Safe Drinking Water Act

This act regulates the protection of aquifers and is designed to ensure the maintenance of safe drinking water. Just as CWA protects surface waters, this act protects ground waters against the injection of wastes and toxics.

Surface Mining Control and Reclamation Act

This act establishes environmental standards for all surface-mining operations. It is managed by the Department of Interior and governs the restoration of land used for surface mining to its original condition.

Toxic Substances Control Act (TSCA)

This statute addresses the control of chemical substances and mixtures whose manufacture, processing, distribution, use, or disposal present potential risks to people or the environment. Included are requirements for testing chemical substances and mixtures, for establishing inventories of commercial chemicals, for notification prior to the manufacture or use of new chemicals, and for the reporting of information about potential hazards that might be caused from the manufacture, use, storage, or transportation of chemical substances or mixtures.

International Laws

Information about the environmental laws of other nations can be obtained from the sources listed below. These organizations are listed with their addresses in Section II.

- International Chamber of Commerce
- UN Environment Programme Industry & Environment Office
- Agency for International Development (AID)
- SustainAbility, Ltd.
- United Nations Development Programme
- United Nations Industrial Development Organization
- World Commission on Environment and Development

Every country has a variety of organizations that can serve as resources to businesses about the law and environmental performance. In Britain, for example, there are organizations such as the following:

- Centre for International Law, Kings College London, Manresa Road, London SW3 6LX
- Department of Trade and Industry, Warren Spring Laboratory, Gunnels Road, Stevenage, Herts SG1 2BX
- Confederation of Business Industry, Centre Point, 103 New Oxford Street, London WC1A 1DU
- Environmental Law Foundation, Rubinstein, Callingham, Polden and Gale, 2 Raymond Buildings, Gray's Inn, London WC1R 5BZ
- Association for the Conservation of Energy, 9 Sherlock Mews, London W1M 3RH
- Association for Environmentally Conscious Building, Wiondlake House, The Pump Field, Coaly, GL11 5DX
- British Plastics Federation, 5 Belgrave Square, London SW1X 8PD
- Environment Council, 80 York Way, London N1 9AG
- Environmental Transportation Association, 15a George Street, Croydon CRO 1LA
- Industrial Council for Packaging and the Environment, 10 Greycoat Place, London SW 1P

SECTION IV: JOURNALS AND MAGAZINES

The number of different publications in the environmental field is expanding at a very fast pace. Listed below are only the ones that I have had an opportunity to consult and which I found to have information relative to sustainable performance.

Amicus Journal
Automotive Industries

Biocycle
Bioscience
Buildings
Business and Society Review
Business and the Environment
Business Ethics
Business Insurance
Chemical and Engineering News
Chemical Engineering Progress
Chemical Week
Chemistry and Industry
Community Development Journal
The Conservationist
Earth
Earthwatch
The Ecologist
Environment
Environmental Science and Technology
EPA Journal
Garbage
Industry and Environment
Iron and Steel Engineer
Journal of the Air & Waste Management Association
The Journal of American Enterprise
The Journal of Commercial Bank Lending
Management Review
Mergers and Acquisitions
Metal Finishing
Mining Engineering
Mother Earth News
Mother Jones
The Nation's Business
Nature
Occupational Health and Safety

Oil & Gas Journal
Omni
Paper Trade Journal
Plant Engineering
Pollution Engineering
Power
Progressive Grocer
Pulp & Paper
Resources
Science
Scientific American
Tappi Journal
Technology Review
Water and Pollution Control
Water Environment & Technology
Worldwatch

Section V: Environmental-Audit Checklists

Environmental-audit checklists are instruments designed to serve the purposes of the audit. The content of checklists will vary with the kind of function or facility being audited; the particular environmental performance being audited; and whether the audit is to determine compliance, opportunities to go beyond compliance, or risk. I have included below examples of the following kinds of audit checklists:

1. A water-pollution checklist to audit compliance with regulations.
2. A checklist to identify opportunities to improve the use of company vehicles.
3. A checklist to determine possible risk in land or facilities acquisition.

Water Pollution Checklist

This is a partial and incomplete list and does not represent the full range of items that should be audited.

REGULATIONS	ITEM	COMMENTS	FINDING
Clean Water Act			
	Quantity of oil discharge into surface waters meets standards.		
	Notifications of oil discharges are properly made to National Response Center.		
	Spill-prevention control and countermeasure plan is in place.		
	Spill-prevention control and countermeasure plan meets regulatory requirements.		
	Facility meets all standards relative to pretreatment of pollutants discharged.		
	Ships do not discharge any noxious pollutants or oil into navigable waters.		
	Marine-oil transfer facility is operating in compliance with regulations.		
	Oil- and hazardous-material transfer operations are in compliance.		

Checklist to Reduce Energy Consumption of Company Vehicles

This is a partial and incomplete list and does not represent the full range of items that should be audited.

GOAL	ITEM	COMMENTS	RESULTS
To reduce total energy costs associated with use of company automobiles			
	Company policy issued.		
	Bicycles provided for employee use.		
	Optimum maintenance cycle established.		
	Maintenance cycle followed.		
	Maintenance records up to date.		
	Optimum speed determined for minimum gasoline consumption.		
	Employee drivers have received driver-efficiency training.		
	Criteria established for purchasing new vehicles to maximize mpg.		

Checklist to Assess Risk in Land or Facilities Purchase

QUESTION	ITEM	COMMENTS	FINDING
What liabilities may convey with the property?			
	What has been previous use of the land?		
	Are ground strata free from evidence of previous dumping?		
	What is the depth of the water table and its vulnerability?		
	Are there underground liquid-storage tanks?		
	Is there a record of previous fines?		
	What are the projected costs for required cleanups?		
	What is the perception of the community concerning proposed use of the land?		

Section VI: The Sustainable-Performance Assessment (SPA)

The Sustainable-Performance Assessment (SPA) measures the degree to which an organization is responding to the environmental challenge and positioning itself to stay competitive in the environmental age.

Response Levels

The SPA is based on the assumption that we can distinguish at least four distinct levels at which an organization will function as it responds to environmental pressures. These levels are as follows:

1. Compliance with the law
2. Nonintegrated initiatives
3. Integrated environmental plan and initiatives
4. Sustainable performance (SP)

For purposes of assessment, a "No Response" level has been included in the SPA. This level means exactly what the title suggests, i.e., there is no evidence that (relative to the criteria used in the SPA) the organization has made any response.

Level One: Compliance

The first stage of responsiveness to the environmental challenge is largely concerned with meeting the demands of the law and regulatory requirements. At this stage, the organization has not become sensitive to many of the other pressures created by the multiple needs of its stakeholders, an aroused citizenry, green consumers, and the threat posed by greener competitors.

This response level is typified by an organization's beginning to define its current level of compliance and its level of risk. The goal is minimum compliance with minimum risk at minimum cost.

At this stage, there are no proactive environmental policies or specific improvement objectives that go beyond compliance.

Level Two: Nonintegrated

At this stage, organizations begin to go beyond the strict requirements of the law and undertake various kinds of special environmental programs—beginning to anticipate potential areas of future liability, reducing the use of energy, taking advantage of obvious opportunities to reduce waste and packaging, emphasizing conservation and good housekeeping, etc.

Organizations at this stage begin to anticipate new developments in environmental regulations. Audits are used, but the focus still is largely on compliance. There is no unifying policy for SP, and there are no integrated improvement objectives.

Level Three: Integrated

This is the first level of response at which most of the following initiatives exist at least to some clear and verifiable degree: a published SP policy and specific improvement objectives that go beyond compliance; established baselines for environmental performance and improvement monitored against these baselines; training for SP fully implemented; improvement projects in place; investment made to develop or acquire new environmental technologies; auditing and reporting system being implemented; coalitions exist; and management and human resource systems are revised to support SP.

Level Four: Sustainable Performance

At this level of response, the organization

1. Publishes a policy for SP and specific improvement goals with the full involvement of all its stakeholders and has a plan for revising this policy and these goals to reflect new and more rigorous improvement initiatives.
2. Develops and revises baselines for environmental performance for all key inputs from the environment and all key outputs into the environment (direct and indirect). Maintains data on all processes to reclaim waste and secondary products.
3. Has a well-developed SP training program. Ensures that training for SP is fully integrated into the organization's total training program; that training is provided in new environmental technologies; and that quality time is provided for support and stimulation of learning, acquisition of new technologies, and innovation for SP.
4. Has improvement objectives and projects in place. Creates specific projects to design or revise all services and products to ensure their full compatibility with nature's ecosystems (e.g., specific projects to reduce packaging, closed-loop processing cycles, marketing of wastes as resources for other companies).
5. Has processes to ensure that it supports research and the development of new technologies and uses new technologies to improve its environmental performance.
6. Uses a fully developed and integrated auditing system to assess and track environmental performance. This system includes not only information

on technical performance (compliance, emissions, waste, etc.) but also information on management performance and the performance of all support systems.

7. Initiates and participates in coalitions and partnerships with other industry members, government agencies, professional groups, and the like, to share information, do joint problem solving, develop standards, and support the development of technology.

8. Has made the environment part of all management systems. Information about environmental performance is reported with other key organizational performance data. Environmental performance is part of the overall information-management system. Sustainable performance is understood by all members of the organization to be a job responsibility. All human resource systems have been revised to support SP.

Uses of the SPA

The SPA can be used in the following ways:

1. As an organizational assessment tool. In this case, the whole work force (or an appropriate sample) completes the SPA and the data are used to develop a baseline of the organization's response level.

2. As a management-feedback tool. In this case, managers use the SPA to communicate their own perceptions of how the organization is responding to the environmental challenge.

3. As a third-party auditing tool, to evaluate an organization's level of environmental performance. In this

case, the audit team uses the SPA and from interviews, analysis of documents, direct observation, etc., provides an assessment of the organization's performance.

Directions for Using the SPA

1. The SPA evaluates an organization's performance according to the degree to which it has reached eight milestones that are essential to achieving SP. These milestones are
   - SP policy statement published.
   - SP baselines established.
   - Initial SP training accomplished.
   - Initial set of improvement projects underway.
   - Development of environmental technologies being supported.
   - Auditing and reporting systems functioning.
   - Coalitions exist.
   - Management and human resource systems revised to support SP.
2. Each milestone has seven criteria, denoted "a" through "g." To complete the SPA, rate each criterion from 0 to 4.
   - "0" means that there is no evidence that this criterion exists at all.
   - "1" means that there is very little evidence that this criterion exists.
   - "2" means that there is some clear suggestion that this criterion exists.
   - "3" means that there is no doubt that this criterion exists at least to some degree.

- "4" means that there is no doubt that this criterion fully exists.

3. Place the rating for each criterion in the column headed "Rating." The total score for each milestone is obtained by adding each of the ratings for each criterion. The lowest possible score is 0. The highest possible score for each milestone is 28.

Interpreting the SPA

1. The total rating for each milestone suggests how far along an organization is toward meeting that milestone and at what level the organization is functioning. The following breakdown can be used as a guide:

   - 0–2.5   Compliance Level
   - 2.6–3.0 Nonintegrated Level
   - 3.1–3.6 Integrated Level
   - 3.7–4.0 Sustainable-Performance Level

2. The rating on each milestone and an analysis of the ratings on the criteria for that milestone can be used to develop baselines and improvement projects.

The Sustainable-Performance Assessment

Milestone	Response Level					
	No Response	Compliance	Nonintegrated	Integrated	SP	Rating
1. POLICY						
(a) management demonstrates full support	0	1	2	3	4	
(b) developed with full involvement of work force	0	1	2	3	4	
(c) developed with full involvement of all stakeholders	0	1	2	3	4	
(d) policy fully compatible with the Principles of SP	0	1	2	3	4	
(e) leaves no doubt about organization's valuing of the environment	0	1	2	3	4	
(f) ties business interests of organization to best interests of environment	0	1	2	3	4	
(g) specifically includes concept of sustainability	0	1	2	3	4	
TOTAL						

The Sustainable-Performance Assessment (continued)

Milestone	Response Level					Rating
	No Response	Compliance	Nonintegrated	Integrated	SP	
2. BASELINES						
(a) management uses baselines in deciding on improvement projects	0	1	2	3	4	
(b) developed with full involvement of work force	0	1	2	3	4	
(c) developed with full involvement of all stakeholders	0	1	2	3	4	
(d) measurable	0	1	2	3	4	
(e) established for all inputs into organization	0	1	2	3	4	
(f) established for all outputs from organization	0	1	2	3	4	
(g) established for all improvement initiatives	0	1	2	3	4	
TOTAL						

The Sustainable-Performance Assessment (continued)

Milestone	Response Level					Rating
	No Response	Compliance	Nonintegrated	Integrated	SP	
3. TRAINING						
(a) management actively participates in training	0	1	2	3	4	
(b) developed with full involvement of work force	0	1	2	3	4	
(c) developed with full involvement of all stakeholders	0	1	2	3	4	
(d) includes Principles of SP	0	1	2	3	4	
(e) includes all required SP strategies	0	1	2	3	4	
(f) includes all needed SP tools	0	1	2	3	4	
(g) fully integrated into organization's training system	0	1	2	3	4	
TOTAL						

The Sustainable-Performance Assessment (continued)

Milestone	Response Level					Rating
	No Response	Compliance	Nonintegrated	Integrated	SP	
4. PROJECTS						
(a) senior management held responsible for SP improvement projects	0	1	2	3	4	
(b) fully involve the whole work force	0	1	2	3	4	
(c) fully involve all stakeholders	0	1	2	3	4	
(d) go beyond solving problems to meet compliance requirements	0	1	2	3	4	
(e) tied directly to organization's environmental-performance objectives	0	1	2	3	4	
(f) provide measurable results	0	1	2	3	4	
(g) have potential for improving organization's competitive position	0	1	2	3	4	
TOTAL						

The Sustainable-Performance Assessment (continued)

Milestone	Response Level					
	No Response	Compliance	Nonintegrated	Integrated	SP	Rating
5. TECHNOLOGY						
(a) management demonstrates full support for use of new technology for SP	0	1	2	3	4	
(b) work force fully involved in helping organization use new technology for SP	0	1	2	3	4	
(c) stakeholders fully involved in helping organization use new technology for SP	0	1	2	3	4	
(d) organization actively supports conferences focused on technology for SP	0	1	2	3	4	
(e) organization initiates joint ventures with other organizations to improve technology for SP	0	1	2	3	4	
(f) organization maintains in-house information resources on technology for SP	0	1	2	3	4	
(g) work force has easy access to training in technology for SP	0	1	2	3	4	
TOTAL						

The Sustainable-Performance Assessment (continued)

Milestone	Response Level					Rating
	No Response	Compliance	Nonintegrated	Integrated	SP	
5. AUDITING/REPORTING						
(a) senior management directly involved in regular auditing and reporting of environmental performance	0	1	2	3	4	
(b) work force fully involved in regular auditing of environmental performance	0	1	2	3	4	
(c) stakeholders fully involved in regular auditing of environmental performance	0	1	2	3	4	
(d) auditing exists for all major organizational elements (facilities, production, engineering, marketing, logistics, etc.)	0	1	2	3	4	
(e) results of audits systematically reported to all stakeholders	0	1	2	3	4	
(f) auditing develops data that goes beyond information on compliance	0	1	2	3	4	
(g) results from audits are turned into improvement projects	0	1	2	3	4	
TOTAL						

The Sustainable-Performance Assessment (continued)

Milestone	Response Level				
	No Response	Compliance	Nonintegrated	Integrated	SP	Rating
7. COALITIONS						
(a) management participates in coalitions with other organizations and groups concerned with SP	0	1	2	3	4	
(b) work force fully involved in coalitions with other organizations and groups concerned with SP	0	1	2	3	4	
(c) organization fully involved with stakeholders to build coalitions concerned with SP	0	1	2	3	4	
(d) organization participates in coalitions with local, state, and federal governments	0	1	2	3	4	
(e) organization participates in coalitions with representatives of own type of business	0	1	2	3	4	
(f) organization participates in coalitions representing national and international business interests	0	1	2	3	4	
(g) organization participates in coalitions with environmentalists	0	1	2	3	4	
TOTAL						

Appendix / 321

The Sustainable-Performance Assessment (continued)

Milestone	No Response	Compliance	Nonintegrated	Integrated	SP	Rating
8. SYSTEMS SUPPORT						
(a) concern for SP has positive influence on all management decisions	0	1	2	3	4	
(b) work force fully aware of responsibility for SP	0	1	2	3	4	
(c) environmental concerns of stakeholders fully considered in all organizational decisions	0	1	2	3	4	
(d) systems exist for rewarding work that improves the organization's environmental performance	0	1	2	3	4	
(e) accounting systems include information on costs relative to environmental performance (e.g., credits and debits relative to wastes, emissions, reclamation of materials, energy)	0	1	2	3	4	
(f) accountability exists for every organizational member for environmental performance	0	1	2	3	4	
(g) environmental performance of people considered in all personnel transactions (e.g., hiring, promoting)	0	1	2	3	4	
TOTAL						

COMPANY/ORGANIZATION INDEX

A

ABB Flakt, 2
Abt Associates, 2
Adolph Coors Company, 4
Agency for International Development, 295, 301
Agricultural Resources Center, 295
Alberto-Culver Company, 103
Alcoa Aluminum, 250
  recycling efforts by, 231
Alliance for Environmental Education, 88
Alliance to Save Energy, 295
Allied Chemical, 234
Allied-Signal, Inc., 87, 255
  policy statement of, 51
American Council for Energy-Efficient Economy, 296
American Cyanamid, 234
American Enviro Products, Inc., 103
American Express, 276
American Productivity and Quality Center, 262
American Talk Issues Foundation (ATIF), 86
American Textile Manufacturers Institute (ATMI), 90
Amoco Corporation, 2, 90
Anheuser-Busch, 39, 110
  employee involvement at, 180
  reclamation efforts by, 228–229
Apple Computer, 45
ARCO Corporation, 233–234
AT&T, 2, 39, 110, 208
  benchmarking by, 263–264
  environmental objectives of, 58
  materials reduction by, 232
  product replacement by, 227–228
  recycling efforts by, 229
Automotive Consortium on Recycling and Disposal (ACORD), 93–94

B

BankAmerica, 94
Baxi Heating, 18–19
BMW, 93
Body Shops International, 39
  environmental activism by, 207
  recycling efforts by, 232
Boeing Corporation, 60
Borden, Inc., 275
Brio Refining, Inc., 80
British Gas, 2, 60, 108
British Petroleum, 60, 108, 111, 234
Burlington Resources, 98
Business Council for Sustainable Development (BCSD), 8, 95, 108, 121, 184, 294

C

California Edison, reclamation efforts by, 228
Canadian Chemical Producers' Association, 171
Chemical Manufacturers Association, 91, 171
Chevron Oil Company, 16, 60, 88, 90, 94, 208, 250
  SMART program of, 274
  system replacement by, 227
Chrysler Corporation, 94
Ciba-Geigy, recycling efforts by, 230–231
Clean Sites, Inc., 296
Coalition for Environmentally Responsible Economies (CERES), 8, 25, 38, 60, 75, 76, 95, 108, 121, 122, 184, 286, 296
Coalition of Northeastern Governors (CONEG), 91–92
Coca-Cola Company, recycling efforts by, 231

Computer and Business Equipment Manufacturers Association, 91, 184
Conference of European Rectors, 88
Consumer Product Safety Commission, 299
Corporate Environmental Advisory Council, 46
Council for Economic Priorities, 38
Council for Solid Waste Solutions (CSWS), 102
Council on Plastics and Packaging in the Environment, 266
Cummins Engine Company, 105

D

Dap, Inc., 89
Data General, waste reduction by, 234
Del Monte, waste reduction by, 234
Deloitte & Touche, 3
Digital Equipment, 15
Dow Chemical, 46, 90, 108, 111, 208, 264
  environmental concerns of, 277
  policy statement of, 50
  recycling efforts by, 229
  WRAP program of, 274
DuPont Corporation, 16, 39, 60, 90, 108, 250, 264
  movement toward environmental responsibility, 113
  packaging innovations by, 146
  pressure on, 83, 84

E

ECO Rating International, 96
EcoSource, 110
Edison Electric Institute, 88
Elmwood Institute, 296
El-Nasr Glass and Crystal Company, 14
Energen, 98
Environmental Defense Fund, 38, 84, 296
Environmental Law Institute, 296

Environmental Literacy Institute, The (TELI), 87
Environmental Protection Agency (EPA), 18, 78, 79–80, 82, 89, 91, 104, 126, 127, 138, 148–149, 200, 255, 296, 298, 299
  Air Toxics Reduction Program of, 275
  audit list proposed by, 252–253
European Community (EC), 106
Exxon Corporation, 2, 80, 200

F

Federal Paper Board, 234
Food Marketing Institute, 99
Ford Motor Company, 94
  reclamation efforts by, 228
Ford Rover, 93
Fuji Photo Film U.S.A., Inc., 88

G

General Agreement on Trade and Tariffs (GATT), 294
General Dynamics, process replacement by, 227
General Electric Company
  POWER program of, 274
  waste recovery by, 147
  waste reduction by, 234
General Motors, 94, 105, 281
Geneva Steel, 17
George Lithograph, 94
GE Plastics, 90
German Environmental Management Association (B.A.U.M.), 88, 274
Global Environmental Management Initiative (GEMI), 38, 60, 95–96, 108, 110, 121, 184, 261–262, 296
Global Tomorrow Coalition, 108
Golf Course Superintendents Association of America (GCSAA), 276
Golin/Harris, 101
Green Cross Certification Company, 100, 296
Greenpeace, 84, 104

Green Seal, Inc., 100, 296
Gulf Coast Acid Team, 229

H

H. B. Fuller, 264
Heinz Company, 110
Henry Ford Hospital, recycling efforts by, 229
Hewlett-Packard Company, 15, 39, 45, 250
  energy reduction by, 232
Hyatt Hotel and Resorts, 15

I

IBM (International Business Machines), 39, 111, 250
  energy management by, 231–232
  recycling efforts by, 229
IMC Corporation, 111
Imperial Chemical Industries, 274
Industry Cooperative for Ozone Layer Protection (ICOLP), 184
Inkwell of America, 99
Institute for Alternative Agriculture, 296
Institute of Food Technologists (IFT), 90, 184
Institute of Organizational and Social Studies, 201
Institute of Packaging Professionals (IoPP), 45, 91, 184
Intel, 263–264
International Alliance for Sustainable Agriculture, 296
International Benchmarking Clearinghouse, 262
International Chamber of Commerce, 3, 8, 10, 25, 38, 60, 75, 95, 108, 121, 122, 277, 289, 294, 301
  definition of auditing, 61–62
  definition of environmental auditing, 181
  on the payoffs of auditing, 241–242
International Council of Scientific Unions (ICSU), 92

International Environmental Bureau, 3, 294
International Environmental Information System, 294
International Hotel Association, 276
International Institute for Sustainable Development (IISD), 8, 121
International Network for Environmental Management, 3, 294
Investor Responsibility Research Center (IRRC), 96, 97, 177, 296

J

J. Walter Thomson, 99
Jaguar, 93
Joint Appeal by Science and Religion for the Environment, 93

K

Kansai Electric Power, 126
Kansas Power and Light Company, 275

L

Lister Butler, 100
Loews Hotels, 276
Low Emissions Technologies R&D Partnership, 94

M

Management Institute for Environment and Business (MEB), 39, 60, 87, 121–122, 184, 297
Marriott Hotels, 276
McDonald's, 111
  policy statement of, 51
  pressure on, 83, 102
McKinsey & Company, 112
Minnesota Mining and Manufacturing (3M), 111, 208, 250, 264
  3Ps program of, 273–274
  energy conservation by, 54
  waste elimination by, 142
Mobil Corporation, 104
Monsanto Company, 80, 90
  recycling efforts by, 230

N

National Audubon Society, 84, 88
National Environmental Association, 110, 262
National Recycling Coalition, 110, 263
National Science Teachers Association, 88
National Solid Wastes Management Association, 297
National Wildlife Federation, 84, 88, 111, 263, 297
Natural Resources Defense Council, 39, 84
Nature Conservancy, 276
Nauerz Company, 228
New Alternatives, 98
Nippon Steel, 60
Nissan, 93
Nordic Council of Ministers, 100
Norsk Hydro, 48–49, 60, 250
Northern Telecom, Inc., 16–17, 39, 208
    policy statement of, 49–50

O

Occupational Safety and Health Administration (OSHA), 82
Orange and Rockland Utilities, Inc., 275
Organization for Economic Cooperation and Development, 294
Organizations for Economic Cooperation, 114

P

Pacific Bell, 94
Pacific Gas & Electric, 94
Partnership for Plastics Progress, 90, 184
Patagonia Company, materials reduction by, 232
Pax World, 98
Peugeot, 281
Peugeot Talbot, 93

Proctor & Gamble, 39, 60, 108, 250
    communication efforts by, 179
    composting efforts by, 149

Q

Quantum Chemical, 90

R

Ramada International Hotels and Resorts, 276
Rand Corporation, 80
Recycled Paper Coalition, 94
Renew America, 297
Resources for the Future, 297
Responsible Care Program, 108, 171–172, 278, 297
Rexham Corporation, 16
    reclamation efforts by, 147
Riker Laboratories, pollution control by, 230
Robert Wood Johnson Medical School, 48
Rockwell International Corporation, 80
Rocky Mountain Institute, 126, 297
Rohm and Hass, 2, 111
Royal Supply, 234

S

Safeway, 94
Sears, Roebuck and Company, materials reduction by, 233
Sharp Company, 126
Shell International, 2
Siemens, 281
Sierra Club, 84
Soil and Water Conservation Society, 297
Solid Waste Composting Council, 149
STORA, 274
Sun Mycrosystems, 45
SustainAbility, Ltd., 2, 8, 39, 294, 301
    research work by, 101–102, 107

T

Toyota Motor Sales, Inc., 88
Triple A Machine Shop, Inc., 79
Tufts University Center for Environmental Management, 8, 297
Turner Broadcasting System (TBS), 86

U

Union Carbide Chemical and Plastics Company, 80, 87, 255
Union of Concerned Scientists, 297
United Airlines, 231
United Nations Center for Transnational Corporations, 95, 295
United Nation Conference on Environment and Development, 1, 107
United Nations Development Programme, 295, 301
United Nations Economic Commission for Europe, 295
United Nations Environment Programme, 8, 54, 60, 108, 295
United Nations Industrial Development Organization, 39, 295
United Nations World Commission on Environment and Development (WCED), 39, 76, 120, 301
United States Chamber of Commerce, 78
United States Environmental Agency, 111, 263
United States Federal Trade Commission, 103
United States Office of Technology Assessment, 226
United States Postal Service, 276–277
United Technologies, Inc., 79
University of California at Irvine Medical Center, 228
U.S. Industries, pollution control by, 230

V

Valvoline, 110
Vauxhall, 93

W

Wallace Computer Services, 94
Warner Center Association Recycling Program, 234
Waste Management, Inc., 81
Westin Hotels and Resorts, 276
Weyerhaeuser Corporation, 234
White House Council on Environmental Quality, 111
Wilderness Society, 84
Wisconsin Electric Power Company, 111
World Environment Center, 108, 111, 263, 295
World Resources Institute, 108, 295
Worldwatch Institute, 295
World Wide Fund for Nature, 108

X

Xerox Corporation, 264
    recycling efforts by, 231

SUBJECT INDEX

A

AaKvaag, Torvild, 49
Accountability, for environmental performance, 64
Accounting information, reliance on, 202
Acid rain, 107, 176
Action
 in the audit report, 259
 as a leadership responsibility, 280–281
 in the reporting process, 246–247
Activist organizations, 74–75, 84–85
Administrator's Award, 111
Advertising, green, 25, 103–104
Advisory groups, functions of, 46
Africa, poverty in, 129
Agriculture
 pollutants related to, 176
 practices involved in, 123
 system of, 133
Air pollution, 68, 73, 107
Ammonia, recovering, 147
Annual reports, environmental information in, 275
Annual Reports Abstract, 263
Aroused citizenry, 75, 85–89, 99, 113
Assessment, of sustainable performance response level, 62. *See also* Auditing
Associations
 environmental, 89–94
 pressure from, 76
Atmosphere, pollution of, 107
Audit checklists, environmental, 304–307
Auditing programs, 239–240
 developing, 240–251
 payoffs of, 241–242
 questions for, 243–245
 review and revision of, 250–251
Auditing system, 61–62

Audit manual, 258
Audit reports, 259–261
Audits, 31, 239. *See also* Environmental audits
 conducting, 255–261
 defining the scope of, 256
 preparing for, 257–258
 purposes of, 62
 reporting findings and recommendations of, 259–261
 of stakeholder needs, 47
 types of, 254
Audit team, establishing, 256–257
Automotive industry
 coalitions in, 94
 use of paints in, 274
Awards
 environmental, 110–112, 276
 for quality, 262–263
Awareness, management, 197–199

B

Baselines
 developing, 219, 262
 establishing, 52–55
Benchmarking, 31, 52, 261–264
Biodiversity, 68
Bioenergy, recovery of, 228–229
Biological agents, 52
Biosphere, protection of, 287
Blanchard, Elwood, 84
Blue Angel labeling, 100
"Bottom line" benefits, of environmental action, 109–110
Britain, environmental coalitions in, 94
British Gas, 2
Brundtland Report, 120
Business
 as an analogue to ecosystems, 166
 environment friendly, 1
 greening of, 2 3, 272

quality as a goal of, 130
relationship to environment, 4
response to environmental problems, 68–71
Business and Industry News, 263
Business Charter for Sustainable Development, 8, 10, 95, 122, 290–293
Business organizations, environmental, 90–91
Business partnerships, 184
Business success, relationship to environmental performance, 50
Buying habits, change in, 99. *See also* Consumer preference

C

California, Toxic Law in, 78
Canada
  ecolabeling in, 99–100
  waste management in, 55
Carbon-dioxide emissions, 138, 148, 165–166, 176
Carson, Rachel, 71, 165
Cement industry, 54
Centers for environmental research and education, 87
CEO culpability, 82
CERES Principles, 8, 10, 95, 122, 287–289
Change, pressure for, 74–77
Checklists, 259, 260, 304–307
Chemical industry
  cooperation in, 171
  energy reduction by, 54
  noncompliance in, 80–81
  product stewardship by, 91
  response to environmental concerns, 277–278
Chemicals
  costs of disposing of, 227
  discharge of, 148
  reuse of, 230–231
Chernobyl meltdown, 86
China, energy reduction in, 14
Chlorofluorocarbons (CFCs), 84
  cooperation in dealing with, 184
  creation of, 165
  increased concentration of, 176
  Montreal Protocol and, 107
  reducing the use of, 227–228
Citizenry
  arousal of, 75, 85–89, 99
  education of, 89
Clean Air Act (CAA), 77, 81, 298
  violations of, 82
Clean-air standards, 77–78
Clean Water Act, 77, 299
Climatic change, 148, 166. *See also* Global warming
Closed-loop processing (CLP), 18, 234, 268
Coalitions. *See also* Organizations
  environmental, 91–93
  industrial, 93–94
  international, 171–172
  participation in, 220
  political, 93
  pressure from, 75
  role in supporting environmental technologies, 63
Cogeneration, 54
Commitment
  of management, 39–42
  securing, 6
Communication. *See also* Information
  characteristics of, 177–179
  with stakeholders, 250
Community building, 157
  sustainable performance and, 170–174
Community Right-to-Know Act, 77, 83, 177, 298
Competencies
  general, 56
  learning of, 41
  organizational, 194, 195
  relationship to influence, 42–44
  required by leadership, 221, 223–224
  sustainable performance and, 37–38, 210–213
Competition
  global, 106

Subject Index / 331

"going green" and, 2–4
pressure from, 76, 108–110
Compliance
  auditing and, 239
  challenge of, 2
  characteristics of, 214–215
  government pressure for, 74, 77–79
  legislative, 214, 215
  as a level of response, 215–216
  of multinational corporations, 105–106
  requirements of, 78
Compliance Index, 97
Composting, 148, 149
Comprehensive Environmental Response, Compensation, and Liability Act (CERCLA), 77, 83, 298
Conferences, becoming active in, 60
Congressional Budget Office, 138
Connectedness, in systems thinking, 163–166
Conservation
  of energy, 53–54, 126
  practicing, 224, 225–227
Consultants, sustainable performance and, 9
Consumer-electronics industry, 201
Consumerism, green, 101–102
Consumer preference
  environmental concern in, 95–104
  for green companies and products, 75
Consumers, informed, 102
Continuous improvement, 130, 243. *See also* Improvement
Convention on Civil Liability for Damages Caused During Carriage of Dangerous Goods by Road, Rail and Inland Navigation Vessels, 107
Cooperation
  among industry groups, 104
  in improving sustainable performance, 183–185
  international, 171–172

Coors, William, 4
"Cornucopia" view of nature, 69–70
Corporate Conservation Council Environmental Achievement Award, 111
Corporate Environmental Profiles Directory, 97
Costs
  of capital, 201–202
  distortion of, 114
  of environmental inputs, 137–138
  of noncompliance, 79–81
  opportunities for saving and avoiding, 127–128, 198
  regulatory, 78–79
Criminal liability, 81 83
Criteria for Sustainable Development Management, 95
Culpability
  of CEOs, 82
  employee, 81–82
  for violating environmental laws, 74
Customer
  analyzing the needs of, 150
  role in the profit formula, 198
  in the Systems Model for Sustainable Performance, 149–150
Cycles, environmental, 135

D

Data collection, 245, 259
DDT, 73, 165
Deck, Michael, 97
Declaration of the Business Council for Sustainable Development, 95, 122
Deep-well injection, 52
Designing-for-environment (DFE), 18
Details, paying attention to, 225–227
Developing countries, 25–26, 128–130
Direct outputs, 148–149
Dow Jones News, 263

E

Earth Summit (1992), 1, 107
Ecolabeling, 99–100
  in the European Community, 106
Ecological impacts, of suppliers, 135–136
Ecological reporting, 138
"Ecologo" labeling, 100
Ecology, in the popular consciousness, 71
"Eco-Mark" labeling, 100
Economic well-being, access to, 124
Ecosystems
  complex changes in, 166
  damage to, 121
  defined, 71–73
  relationships among, 125
  self-regulation of, 72–73
  services provided by, 72, 134
EDEX evaluation process, 100
Education
  about environmental problems, 86–89
  for environmental awareness, 86–89
  of management, 113
Educators, sustainable performance and, 9
Elkington, John, 8, 99, 101
Emissions, 54. *See also* Fugitive emissions
  carbon-dioxide, 138, 148, 165, 176
  control of, 94
  national boundaries and, 107
Emissions Efficiency Index, 97
Employees
  culpability of, 81–82
  as a pressure for environmental change, 112–113
  as problem solvers, 212–213
  whistleblowing by, 89
Empowerment
  of employees, 113
  meanings of, 180–181
  regulation and, 210
End-of-pipe assessment, 265
End-of-pipe management, 134–135, 144, 148, 162–163, 203
  costs of, 230
Energy
  conservation of, 53–54, 126–127
  management of, 231–232
  nonrenewable sources of, 123
  reducing the costs of, 228
Energy Star Computer Program, 91
Environment
  as a business partner, 8
  as a customer, 133, 272
  cooperation with, 156
  direct outputs to, 148–149
  impact of processes on, 142
  input/output cycle of, 135
  as a priority, 41
  in the Systems Model for Sustainable Performance, 133
Environmental activist organizations, 84–85
Environmental-audit checklists, 304–307
Environmental auditing, 181–183. *See also* Auditing programs
  requirements for, 253
  types of, 254
Environmental audits, 252–255. *See also* Audits
Environmental awards, 110–111
Environmental awareness, 4
"Environmental" business, violations in, 81
Environmental business networks, 3
Environmental catastrophes, 86, 200
  costs of, 80–81
Environmental challenge, proactive response to, 5
Environmental change
  impacts of, 68
  threats associated with, 67–68
Environmental Choice Program (Canada), 100
Environmental concerns, surveys of, 100–102
Environmental degradation, costs of, 76

Environmental Initiatives for Advertising Agencies, 90
Environmental inputs, 137–139
Environmental issues, proactive stance toward, 2
Environmental laws, 77–79, 297–302
  violating, 74
Environmental Marketing and Claims Act, 104
Environmental performance
  accountability for, 64
  gauging, 52
  integration of, 185–187
  international codes for, 75, 94–96
  tracking, 219–220
Environmental planning, sustainable performance and, 6
Environmental policy, of business and trade organizations, 90–91
Environmental problems
  changes in, 73
  fragmentation of, 70–71
  information and education about, 86–89
  recognition of, 73
  as systematic and global, 71–73
Environmental protection, in developing countries, 129
Environmental publications, 302–304
Environmental Self-Assessment Program, 95–96
Environmental technologies, support for development of, 59–61
Environmental technology teams, 60
Epstein, Marc, 98
Equity, as an element of sustainable development, 120, 124
Estimates, role in waste management, 55
Executives, commitment to sustainable performance, 39–42. *See also* Leadership; Management
Expectation, as a motive, 205, 208–210
Extinctions, 166

F

Federal Hazardous Substances Act (FHSA), 299
Federal Insecticide, Fungicide, and Rodenticide Act (FIFRA), 77, 299
Federal Water Pollution Control Act (FWPCA), 299
Feedback, 20–21, 104
  negative, 137, 163
Fines, punitive, 74, 79–81
Food chain
  changes in, 165
  threats to, 176
Fortune 200 firms, environmental issues in, 2
Fossil fuel, 123
  combustion of, 175–176
  taxes on, 138
Fragmentation, of environmental problems, 70–71
Fraud, in green marketing, 103–104
Fugitive emissions, 52, 53, 244
Full-cost pricing, 76, 112, 114–115

G

Germany
  ecolabeling in, 100
  packaging laws in, 105–106
  waste management in, 55
Global community, development of, 170
Global Forum of Spiritual and Parliamentary Leaders, 93
Global markets, 76, 105–106
Global politics, 107–108
  pressure from, 76
Global threat, response to, 68–71
Global warming, 73, 148, 175
Goals
  of business, 130
  performance, 6
"Going green" movement, 2
  advantages of, 5
  competition and, 3–4
Gold Medal Award, 114
Golf Course Management, 276

"Good Neighbor" program, 276
Government. *See also* Regulations
　pressure from, 74, 77–79
　role in environmental education, 89.
Grazing, effects of, 114
"Green architecture," 18
Green Business Guide, The, 101
Green companies, investment in, 96–98
Green Consumer, The, 99, 101,
Green Consumer's Supermarket Shopping Guide, 99, 101
Greenhouse effect, 17, 176. *See also* Global warming
"Green Lights" program, 126
Green marketing, 4, 103–104
Green market niches, 32, 218, 225, 233–234
Green products, consumer preference for, 75
Green Report II, 103
Green rules, playing by, 277–278
Greenworld Survey, 2, 107
Ground-water contamination, 68, 71, 73, 78

H

Hailes, Julia, 99
Handbook for Environmentally Responsible Packaging in the Electronics Industry, 45
Hazardous Materials Transportation Act (HMTA), 299–300
Health
　effect of ozone depletion on, 176
　threats to, 67–68
Heavy metals, 176
Hiring, for environmental skills, 65
Hotel industry, 15
"Hotels of the New Wave" program, 276
HR 5305, 82–83
Human beings, relationship to environment, 134

Human resource systems
　revision of, 64–65
　sustainable performance in, 7

I

Improvement
　developing objectives for, 263
　focus of, 168
　projects for, 57–59
Incineration, of hazardous waste, 52, 149
India, Union Carbide spill in, 80, 86, 200
Industrial coalitions, 93–94
Industrial waste, 124–125
　eliminating, 226
Industry, greening of, 2–3
Influence, relationship to competency, 42–45
Information. *See also* Communication
　about environmental performance, 177
　about environmental problems, 86–89
　organizational channels for, 163
　reporting of, 245–250
Initiatives
　for improving environmental performance, 49
　joint, 63
Ink recycling, 147
Input
　suppliers and, 139–140
　in the Systems Model for Sustainable Performance, 137–140
Integrated environmental plan and initiatives, as a response level, 217
International coalitions, 171–172
International codes, for environmental performance, 75, 94–96
International Environmental Business Codes, 286–293
International environmental laws, 301–302
International industry groups, 184

International organizations
　environmentalism and, 107–108
　pressure from, 76
International resources, 294–295
Investors, environmentally conscious, 75, 96–98. *See also* Stakeholders
Involvement
　of stakeholders, 45–47
　of the work force, 42–45
Iwasaki, Ed, 45

J

Japan
　ecolabeling in, 100
　energy efficiency in, 126–127
　time horizon in, 201, 202
　waste management in, 55
Jaques, Elliot, 201
Journals, environmental, 302–304
Judeo-Christian beliefs, 69
"Just-in-time" supply and inventory systems, 139

K

Keidanren Global Environmental Charter, 95
"Knowing endangerment" provisions, 83
Knowledge, importance to sustainable performance, 212–213. *See also* Communication; Information

L

Landfills, 52, 180
　capacity of, 148–149
　costs of, 229
Laws, 74, 77–78. *See also* Environmental laws
　compliance with, 214, 215–216
　international, 301–302
　packaging, 106
　state, 83
　toxic-materials, 92
　in the United States, 297–301

Leadership. *See also* Management
　action and, 280–281
　competencies for, 37
　fragmented thinking in, 159
　learning and, 198–199
　limits of, 59
　responsibilities of, 278–281
　in sustainable performance, 8–9
　systems thinking and, 163
　understanding of sustainable performance by, 191–192
Leadership tasks, in the Response Model for Sustainable Performance, 194–196
Legislation. *See* Laws
Liability, for environmental damage, 81. *See also* Culpability
Life-cycle analysis (LCA), 33, 162, 264–268
　phases of, 265–266, 267
Life-cycle management, 146
Limits
　as an element of sustainable development, 121, 124
　of nature, 134
　sustainable performance and, 174–177
London Convention on the Prevention of Marine Pollution by the Dumping of Wastes and Other Matter, 107

M

Macro issues, 125
Magazines, environmental, 302–304
Makower, Joel, 99
Management. *See also* Leadership
　auditing program questions for, 243–245
　commitment of, 39–42
　commitment to an auditing program, 240–243
　by processes, 162, 163
　qualitative characteristics of, 39–47

risk taking and, 200
role in auditing program review, 250–251
time horizon of, 201–203
training of, 165
by "walking around," 41
Management awareness, 197–199
Management-education programs, 113
Management systems
revision of, 64–65
role of environment in, 220
Marketing, green, 4, 103–104
Market niches, green, 233–234
Markets, global, 105–106
Match, as a motive, 204, 207
Materials, reducing the use of, 231–233
Measurement
of performance, 246–247
role in improving environmental performance, 53
Meetings, improving, 163
Metal scrap, 52
Mexico, environmental regulation in, 105, 184
Michigan, environmental legislation in, 83
Milestones
derivation of, 38–39
in the Sustainable-Performance Management Model, 37–66
use in sustainable performance, 23–24, 47–65
Mineral resources, 176
Models. *See also* Response Model for Sustainable Performance; Sustainable-Performance Management Model; Systems Model for Sustainable Performance
motivational, 205–206
open-systems, 27
Molded fiber technology (MFT), 147
Montreal Protocol, 107
Motivational Model, 205–206
Motives
organizational, 194, 195
in the Response Model for Sustainable Performance, 203–210
Multicultural work force, 1
Multinational corporations, compliance of, 105–106

N

National Environmental Education Act of 1990, 89
Nationalism, business and, 4
Natural environment, in the Systems Model for Sustainable Performance, 133–135
Nature
perceptions of, 69–70
unacknowledged inputs from, 137
Negative feedback, 137, 163
Network Earth series, 86
New investment, focuses for, 2
New Mexico, environmental law in, 78
New technologies, supporting and using, 59–61
NIMBY (Not in My Back Yard), 85
Nitrogen oxide emissions, 176
Noncompliance, costs of, 79–81. *See also* Compliance
Nongovernmental organizations (NGOs), 84–85
Nonintegrated initiatives, as a response level, 216–217
Nonrenewable resources, 53, 121, 123, 175
"No regrets" policies, 126–127

O

Objectives, for environmental improvement, 58
Occupational Safety and Health Act (OSHA), 300
Organizational performance, environmental issues and, 65
Organizational response levels, 191–222

Organizations
  activist, 74–75, 84–85
  holistic view of, 163
  new environmental rules for, 271–277
  relationship of suppliers to, 135–136
  revised views of, 159–160
  social conscience in, 168
  sustainable performance in, 7–8, 13–19
Our Common Future, 120
Output, 144–147
  direct, 52, 148–149
  indirect, 52
  in the Systems Model for Sustainable Performance, 144–149
Ozone depletion, 73, 137, 165, 176, 184

P

Packaging
  evaluating, 145–146
  laws related to, 106
  reducing, 231–233
Paper, recycling of, 94, 99, 229
Paper industry, 54
Partnerships, for improving sustainable performance, 183–185
PCBs (polychlorinated biphenyls), 68, 253
Performance
  as an element of sustainable performance, 128–131
  measurement of, 246–247
Performance goals, commitment to, 6
Perry, Ronald, 45
Personal culpability, 81–83
Pesticides, 71, 165
Planning, as a responsibility of leadership, 279–280
Plastics
  misunderstanding about, 102
  recycling of, 229
Policy, environmental, 5

Policy statements, 48–52
  criteria for, 49
  functions of, 279–280
Political coalitions, 91–93
Politics
  effect of business on, 4
  global, 107–108
Pollution
  "bottom line" benefits of reducing, 109–110
  exporting of, 105
  management of, 107
  stresses associated with, 176
  taxation of, 114–115
  work-environment, 169
Polyethylene packaging, 102, 146
Popoff, Frank, 277
Population control, 125
Post-consumer waste, 146
Poverty
  eliminating, 125
  responsibility for, 128–130
"Preferred packaging" guidelines, 92
Premanufacturing notice (PMN), 80
President's Environment and Conservation Challenge Awards, 111
Pressures
  for change, 74–77, 194, 195
  of competition, 108–110
  from environmental groups, 83–84
  for responding to environmental challenge, 24–26, 67–115
  in the Response Model for Sustainable Performance, 196
Principles of Sustainable Performance, 7, 10, 28–30, 155–189
  uses of, 158
Problem-solving activities, 60
Processes
  improving, 59
  modifying or replacing, 227–228
  opportunities for improving, 142–143
  in the Systems Model for Sustainable Performance, 140–144

Production
  clean, 226
  phases of, 265–266
Products. *See also* Secondary products
  modifying or replacing, 227–228
  as outputs, 144–146
Professional coalitions, 91–93
Profit
  as an element of sustainable performance, 127–128
  environmental challenge as an opportunity for, 3–4
  environmental responsibility and, 207
  increase of, 130
  as a motive, 205
  in socially responsible investing, 97–98
  traditional view of, 197–198
Project Copernicus, 88
Public arousal, 86–89
Publications
  environmental, 302–304
  on environmental technology, 60–61
  role in communication, 179
  trade, 262
Punitive fines, 74, 79–81

Q

Quality. *See also* Total quality environmental management (TQEM); Total quality management (TQM)
  as a business goal, 130
  awards for, 262–263
  continuous improvement of, 180–181
  environmental, 1, 64, 139–140
  performance and, 128
Quality of life, threats to, 67–68
Questions, importance to the auditing program, 243–245

R

Race to Save the Planet series, 86

Reclaimed waste, 144, 146–147, 228–231
Recommendations, of audits, 259–261
Recycling
  awards for, 110–111
  of chemicals, 229–231
  of packaging, 146
  of paper, 94, 99, 229
  of plastics, 229
  profit from, 147
"Recycling Works," 231
Reduction, Reuse, and Recycling of Protective Packaging (R_3P_2), 45
Regulations
  compliance with, 78, 208
Regulatory requirements, meeting, 251
Religious beliefs, about nature, 69–70
Reorganization, systems thinking and, 164–165
Reporting
  of audit findings and recommendations, 259–261
  developing requirements and processes for, 245–250
  of environmental performance, 181–183
  system of, 61–62
Resource Conservation and Recovery Act (RCA), 77, 300
Resource management, accounting systems for, 64
Resources
  benign use of, 166
  international, 294–295
  limits on, 124, 174
  mental, 180
  nonrenewable, 121, 139, 175
  for sustainable performance, 293–297
  in the United States, 295–297
Response levels, 30–31, 191–192, 213–220, 308–311
  organizational, 191–222
  in the Sustainable-Performance Assessment (SPA), 236

Response Model for Sustainable
Performance, 10, 192–196
  elements of, 194, 209–210
  leadership tasks of, 194–196
  motive as a variable in, 203–210
  pressures in, 196
  response levels in, 213–220
  screens in, 196–203
Results
  developing a view of, 167
  focus on, 170
Return, as a motive, 205, 208. *See also* Profit
Reward systems, 64
Risk assessment, 239, 253. *See also* Auditing
Roddick, Anita, 207
Russell, Paul, 45

S

Safe Drinking Water Act, 77, 83, 300
Sagan, Carl, 93
Scandinavian countries, ecolabeling in, 100
Scientists, recommendations by, 92–93
Screens
  in the Response Model for Sustainable Performance, 196–203
  use of, 194, 195
Secondary products, reclaiming, 228–231
Self-interest, 199–200
Services
  modifying or replacing, 227–228
  as outputs, 144–146
Silent Spring (Carson), 71, 165
Skills, importance to sustainable performance, 212–213
Smog, 68, 176. *See also* Air pollution
Social decisions, effect of business on, 4
Socially responsible investing (SRI), 97-98
Societies, environmental, 89–94. *See also* Coalitions
Soil destruction, 123

Solar energy, 126–127
Solid waste, increase in, 148–149
Solvents, recovering, 147
Soviet Union, steel industry in, 54
Specialized training, 56
Stakeholders
  communication with, 178–179, 247–248
  cooperation among, 172–173
  involvement in the reporting process, 249–250
  involvement of, 45–47
Standard of living, uniform, 129
State law, liability under, 83
Steel industry, 54
Stewardship, as an element of sustainable development, 120–121, 124
Strategies
  for improving environmental performance, 31–32
  for moving toward sustainable performance, 224–235
"Stretch" objectives, 44, 58
Strong, Maurice, 1–2
Styrene packaging, 102
Subsidies
  effects of, 114–115
  to the energy industry, 137
Success
  environmental performance and, 200
  expectations for, 208
Sulphur dioxide emissions, 176, 274
"Sun Day," 86
Superfund, 77, 80, 264, 298
Suppliers
  inputs and, 139–140
  in the Systems Model for Sustainable Performance, 135–136
Surface Mining Control and Reclamation Act, 77, 300
Surveys
  employee-preference, 112–113
  of environmental concerns, 100–102
  on full-cost pricing, 114

Sustainability, concept of, 123
Sustainable development (SD)
   communal nature of, 121, 124–125
   compared to sustainable performance, 123–127
   defined, 120–121
Sustainable performance (SP), 6–8
   characteristics of, 26–27, 127–131
   commitment of management to, 39–42
   communal nature of, 63
   as a community-building process, 170–174
   as a data-based process, 181–183
   defined, 119, 122
   as an ecologically interdependent process, 166
   examples of, 14–19
   knowledge and skills for, 212–213
   as a level of response, 217–220
   as a limiting process, 174–177
   milestones for, 47–65
   as an open process, 177–179
   as a process of continuous improvement, 179–181
   rationale for, 167–168
   resources for, 293–297
   as a results-oriented process, 167–170
   strategies for moving toward, 224–235
   as a technologically dependent process, 183–185
   tools for, 34, 235–268
   as a total organizational process, 185–187
   TQM (total quality management) and, 211–212
   transition to, 64–65
Sustainable-Performance Assessment (SPA), 10, 32–33, 236–239, 308–321
   preparation for using, 238–239
   using, 237, 311–313
Sustainable-performance baselines, establishing, 52–55

Sustainable-Performance Management Model, 7, 9, 10, 13–35, 279
   description of, 22–34
   milestones in, 35, 37–65
   uses of, 21–22
Sustainable-performance policy statement, 48–52
Sustainable-performance training, 55–57
Systemic development, 121, 125
Systems Model for Sustainable Performance, 10, 27–28, 131–151, 240
   characteristics of, 119–151
   life-cycle analysis and, 265–266
Systems thinking, 157, 159–166
   connectedness in, 163–166
   principles and processes of, 160–163

T

TAPESTRY program, 88
Taxation
   of environmental inputs, 138
   of pollution, 114
Team influence, deploying, 44
Teams. *See also* Audit team
   development of, 186–187
   forms of, 181
   importance of, 179–181
Technology
   dependence of sustainable performance on, 183
   support for developing, 59–61
   training for, 219
Terminology, environmental, 103
Time horizon, 201–203
Tools
   leaders' understanding of, 280–281
   for managing environmental performance, 192–193
   for sustainable performance, 32, 235–268
Total quality environmental management (TQEM), 1, 9
   programs for, 273

Total Quality Environmental Management Primer, 96
Total Quality Management: A Systematic Approach to Continuous Environmental Improvement, 179
Total quality management (TQM), 1, 9, 139
 benchmarking in, 261
 change brought by, 162
 improvement projects in, 57
 as a process of improvement, 210–211
 as a survival strategy, 40
 sustainable performance and, 211–212
 training for, 212
 value shifts involved in, 155–156
 weaknesses of, 186
Toxic-materials legislation, 92
Toxic Substances Control Act (TSCA), 77, 78, 80, 83, 301
Toxic waste, 176
Trade and Industry ASAP, 263
Trade organizations
 benchmarking and, 263
 environmental, 90–91
Training
 consistency with sustainable performance, 65
 for environmental technology, 219
 of managers, 162
 specialized, 56
 sustainable-performance, 55–57, 219
Transfer technology, 148

U

Ultraviolet (UV) radiation, 176
Understanding, as a responsibility of leadership, 278–279
Union Carbide spill, 80
United States
 ecolabeling in, 100–101
 environmental coalitions in, 94
 environmental laws in, 297–301
 resources for sustainable performance in, 295–297
University of Florida, environmental education at, 87
Unrecovered waste, 144, 148–150

V

Valdez oil spill, 80, 86
Valdez Principles, 75, 95. *See also* CERES Principles
Values
 necessary shifts in, 155–156
 organizational, 160
Variables, tracking, 244–245

W

Warning to Humanity, 92
Waste
 "bottom line" benefits of reducing, 109–110
 marketing of, 233–234
 reclaimed, 144, 146–147, 228–231
Waste management, 54–55, 144, 148–149
 reducing costs of, 128
Water pollution, 68
Water treatment, 144
Whistleblowing, 89
Woolard, Edgar, 113
Work environment, improving environmental performance in, 168–169
Work-flow process, 141
Work force
 involvement in the reporting process, 248–249
 involvement of, 42–45